Biodiversity and Wind Farms in Portugal

Miguel Mascarenhas · Ana Teresa Marques
Ricardo Ramalho · Dulce Santos
Joana Bernardino · Carlos Fonseca
Editors

Biodiversity and Wind Farms in Portugal

Current Knowledge and Insights for an Integrated Impact Assessment Process

Springer

Editors
Miguel Mascarenhas
Bioinsight
Odivelas, Grande Lisboa
Portugal

Ana Teresa Marques
cE3c—Centro de Ecologia, Evolução e Alterações Ambientais
Faculdade de Ciências da Universidade de Lisboa
Lisbon, Grande Lisboa
Portugal

and

CEABN/InBIO—Centro de Ecologia Aplicada "Professor Baeta Neves", Instituto Superior de Agronomia
Universidade de Lisboa
Lisbon
Portugal

and

CIBIO/InBIO Associate Laboratory
Universidade do Porto
Vairão
Portugal

Ricardo Ramalho
Bioinsight
Cape Town
South Africa

Dulce Santos
Centro de Investigação Interdisciplinar em Sanidade Animal (CIISA), Faculdade de Medicina Veterinária
Universidade de Lisboa
Lisbon
Portugal

Joana Bernardino
CIBIO-InBIO, Department of Biology—Pole of Mitra
University of Évora
Évora, Alto Alentejo
Portugal

Carlos Fonseca
Department of Biology—Group of Adaptation Biology and Ecological Processes
University of Aveiro—CESAM
Aveiro, Baixo Vouga
Portugal

ISBN 978-3-319-60350-6 ISBN 978-3-319-60351-3 (eBook)
https://doi.org/10.1007/978-3-319-60351-3

Library of Congress Control Number: 2017943247

Printed on acid-free paper

This Springer imprint is published by Springer Nature
The registered company is Springer International Publishing AG
The registered company address is: Gewerbestrasse 11, 6330 Cham, Switzerland

Contents

Abbreviations

a.s.l	above sea level
APREN	Associação Portuguesa de Energias Renováveis (in english: Portuguese Renewable Energy Association)
BACI	Before-After-Control-Impact
CHI	Collision Hazard Index
CRM	Collision Risk Models
DEM	Digital Elevation Model
EIA	Environmental Impact Assessment
ES	Environmental Study
EU	European Union
GHG	Greenhouse Gas
GIS	Geographic Information System
GPS	Global Positioning System
GW	Gigawatt
KAI	Kilometric Abundance Index
MW	Megawatt
PNAER	Portuguese National Action Plan for Renewable Energy
PV	Photovoltaic
RAI	Record Abundance Index
SCI	Site of Community Importance
SCIs	Sites of Community Importance
SPAs	Special Protected Areas
USA	United States of America

Chapter 1
Wind Industry in Portugal and Its Impacts on Wildlife: Special Focus on Spatial and Temporal Distribution on Bird and Bat Fatalities

Joana Marques, Sandra Rodrigues, Rita Ferreira and Miguel Mascarenhas

1.1 Introduction

In the quest for climate change mitigation (UNEP 2016), wind is one of the most important and dominant sources of investment in today's energy-demanding world. As per recent international energy statistics, Asia, Europe and North America are responsible for 95% of installed wind energy capacity. However, energy production through wind sources is also starting to expand to other regions, including South America, where an increase of 40% (approximately 3 GW) in installed capacity was registered in 2015 (IRENA 2016). To achieve the European Commission objectives of a minimum of 27% of overall energy production through renewable energy sources by 2030 (UNEP 2016), it is likely that European countries will continue to install wind farms and/or other similar renewable energy production systems, calling for solutions to avoid, mitigate and/or compensate for the negative side effects on biodiversity.

Joana Marques and Sandra Rodrigues: Equal contributors.

J. Marques · S. Rodrigues · R. Ferreira · M. Mascarenhas (✉)
Bioinsight, Odivelas, Portugal
e-mail: miguel.m@bioinsight.pt

J. Marques
e-mail: joana.m@bioinsight.pt

R. Ferreira
e-mail: rita.ferreira84@gmail.com

S. Rodrigues
Department of Statistics and Operational Research,
Faculty of Sciences, University of Lisbon, Lisbon, Portugal
e-mail: sandra.r@bioinsight.pt

M. Mascarenhas et al. (eds.), *Biodiversity and Wind Farms in Portugal*,
https://doi.org/10.1007/978-3-319-60351-3_1

In Portugal, the trend over the last decade has reflected an increase in installed MW, with more than 5 GW installed capacity until 2015 (IRENA 2016). Though this trend is in line with the worldwide expansion of wind energy, the growth of installed power has shrunk since 2011, when the economic crisis affected the country. Wind-generated electricity represented 23% of the energy demand in 2015 in Portugal (e2p 2015), strengthening the contribution of wind energy to achieve the European goals (EU 2009). The current National Action Plan for Renewable Energies (PNAER 2020) adjusts the Portuguese actions for the 2013–2020 period to comply with these goals, increases the quote for electricity production based on renewable energies to 60% (more 5% than the previous plan from 2010), and settles a 400 MW increase in energy production by repowering (PCM 2013).

1.1.1 Portuguese Wind Farm Location in Relation to Natural Values

In 2015, 259 wind farms (with 2367 turbines) were operating in mainland Portugal (the analysis excluded Portuguese islands) (Figs. 1.1 and 1.2, adapted from e2p 2015), all based on-shore (except one) and mainly located in the northern half of the territory. To characterise the Portuguese wind energy and its impacts on wildlife, we analysed information from 44 wind farms monitored between 2005 and 2015, whose reports are publicly available at the national environmental agency ("Agência Portuguesa do Ambiente") and/or both from Bio3 or Bioinsight's internal archive (Table 1.1).

The relative proportion of total (ca. 17%) and sampled wind farms per Portuguese biogeographic region reveals that this dataset represents the variability of biogeographic conditions of wind farms in Portugal (per Costa et al. 1998) (Fig. 1.1). The only sub-sampled bioregions are the Ribatagano-Sadense and the Orensano-Sanabriense Sectors, however, these account only for a marginal 4.3% of the total operational wind farms.

As wind farms are placed in areas of high wind potential, their distribution in continental Portugal is mostly associated with mountain ranges: the average altitude of Portuguese wind farms is 840 m (20–1440 m). Approximately 60% of all wind turbines are over 800 m above sea level (a.s.l.), and 36% are above 1000 m a.s.l. A few are placed in plain areas, mainly near coastal areas to take advantage of maritime wind currents.

Most of the wind turbines are placed in natural or semi-natural areas, including open forests, cleared areas or new forest plantations (25%), scrublands (21%), grasslands (23%) and a smaller portion in farmlands (5%). Potential conflict between wind farms and wildlife arises from the distribution of these infrastructures across natural and semi-natural areas that hold a relatively higher biodiversity. At the national level, for example, 39% of wind farms overlap with the distribution areas of the Iberian Wolf (see Costa et al., Chap. 5); approximately 29% of wind farms are placed at less than 10 km from known bat roosts, though only 6% of these

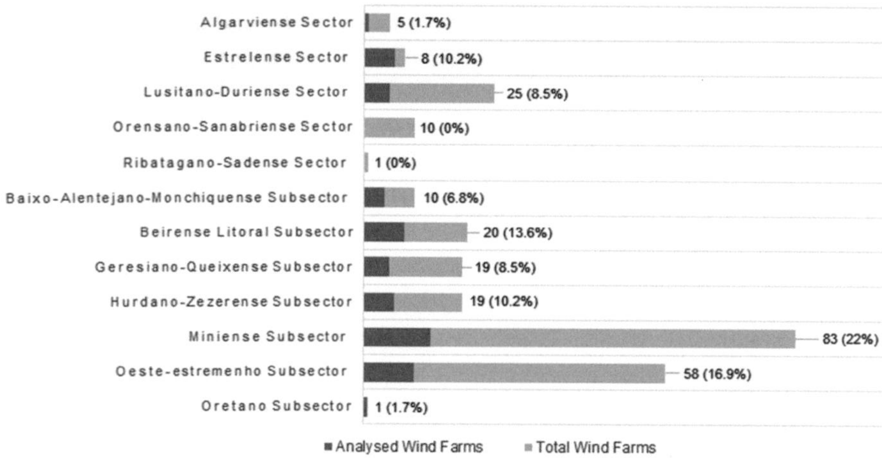

Fig. 1.1 Analysis of the operational wind farms in Portugal in 2015 per biogeographic region: *dark grey bars* represent the number of wind farms considered for analysis in this chapter. The number of reports analysed is indicated and their percentage in relation to the total number of existing farms per biogeographic region is shown in parenthesis; *light grey bars* indicate the total number of wind farms (wind farm locations from e2p 2015)

are considered of national importance (due to the high number of individuals roosting or breeding); six wind farms are located within a well-known migration route for birds (in the southwestern tip of Portugal), and another eight are close to breeding sites of species of conservation concern (particularly eagle and vulture nests at the Douro river escarpments in the north-eastern border and central and southern mountain ranges of Portugal). As birds are particularly prone to collisions while on migration (Lucas et al. 2012a, b; Marques et al. 2014), in the wind farms at this very sensitive area, within the migration route of Sagres, detailed studies were conducted to accurately characterise the migratory bird community and determine the best mitigation measures. In the specific situation, these wind farms are subject to dedicated measures of shutdown on demand using human-assisted radar systems throughout the whole migration period (between mid-July and early-December).

Considering these and other natural values, a National Protected Areas Network was created to safeguard biodiversity from human expansion and associated activities (Decree Law No. 142/2008 of July 31st, amended by the Decree Law No. 242/2015 of October 15th). As a member of the European Union, and as a part of the under the Natura 2000 network, other areas were also classified as Special Protected Areas (SPAs), considered important areas for birds, as adopted by the Directive 2009/147/EC of the European Parliament and of the Council of November 30th; and Sites of Community Importance (SCIs), for example sites with important habitats, as published in the Council Directive 92/43/EEC of May 21st. Both European documents were transposed to the national right through the Decree Law No. 149/99 of April 21st, with posterior amendments by the Decree Law No. 49/2005 of February 24th, and the Decree Law No. 156A/2013 of

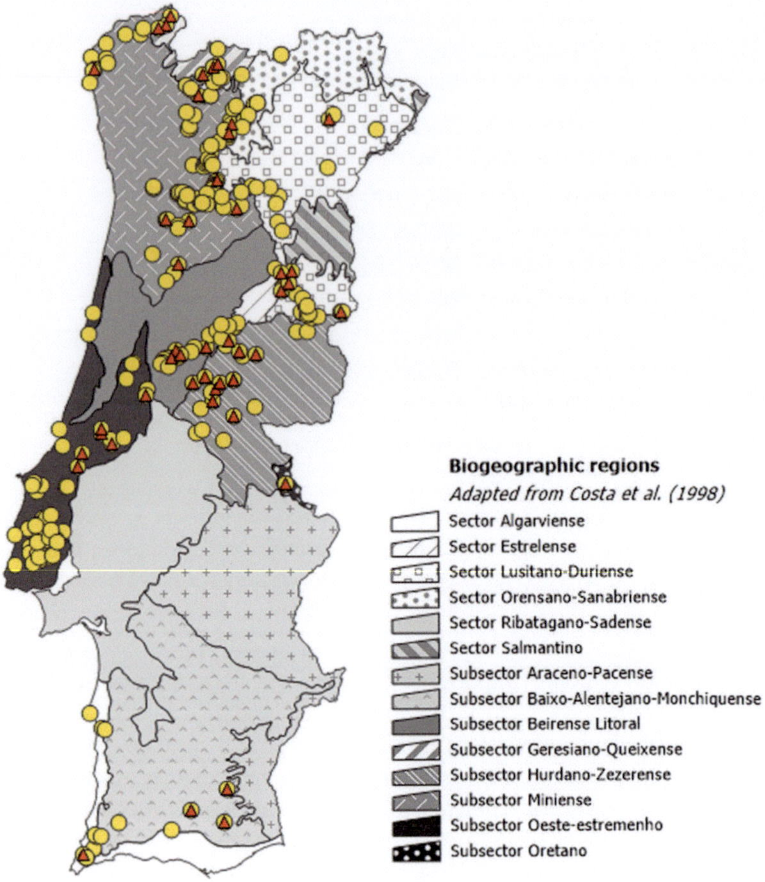

Fig. 1.2 Distribution of wind farms across the biogeographic regions of Portugal (regions adapted from Costa et al. 1998): operational wind farms not included in the analysis (*yellow dots*) and analysed wind farms (*orange triangles*) (analysed data refers to a subsample of datasets from operational wind farms projects between 2005 and 2015) (wind farms locations data from e2p 2015)

November 8th. Approximately 22% of the continental territory is under one of these protection frameworks. Hence it is inevitable that at least some wind farms overlap with these locations: four wind farms are located within National Protected Areas, 71 are within SCIs and two are placed in SPAs (ICNF 2017).

The considerable number of wind farms within SCIs is related to the minor expected impacts on habitats, since the footprint of wind farms is quite small and the layout may be adjusted to avoid priority habitats, being assumed in the Environmental Impact Assessment (EIA) process that this should be sufficient to mitigate negative impacts in areas protected due to their habitat. On the other hand, National Protected Areas and SPAs usually protect faunal values, which are more

Table 1.1 List of wind farms and respective monitoring programme timeframes considered for analysis, including the respective wind farm biogeographic location, installed wind turbines and total power produced (MW)

Wind farm name	Years consulted	Biogeographic region	Number of wind turbines	Produced power (MW)
Açor II	2010–2012	Sector Estrelense	8	16
Alto dos Forninhos	2015	Subsector Oretano	4	8
Alto Minho I	2006–2009	Subsector Geresiano-Queixense	75	150
Alvaiázere	2015	Subsector Oeste-estremenho	9	18
Alvão	2004/2005; 2006	Subsector Miniense	21	42
Barão de São João	2009; 2013	Subsector Oeste-estremenho	25	50
Bustelo e Cinfães	2010–2012	Subsector Miniense	13	26
Cabeço Rainha II	2007; 2009	Subsector Hurdano-Zezerense	15	31.2
Cadafaz II	2010–2011	Subsector Beirense Litoral	9	18
Candal/ Coelheira	2006–2007	Subsector Miniense	20	40
Caramulo	2005–2008	Subsector Miniense	45	90
Carreço-Outeiro II	2008–2009	Subsector Miniense	6	12
Casais	2007–2009	Subsector Miniense	1	2
Chão Falcão (I; II; III)	2005–2009	Subsector Oeste-estremenho	35	80.5
Chiqueiro	2015	Subsector Hurdano-Zezerense	2	4
Fonte da Quelha e do Alto do Talefe	2004–2006	Subsector Miniense	18	27
Gardunha	2006–2010	Subsector Hurdano-Zezerense	57	114
Guarda	2007–2009	Sector LUSITANO-DURIENSE	8	4
Lagoa Funda	2010	Sector Algarviense	6	12
Lousã (I; II)	2004–2010	Subsector Beirense Litoral	34	85
Malhanito	2015	Subsector Baixo-Alentejano-Monchiquense	29	58
Meadas	2015	Subsector Miniense	3	9

(continued)

Table 1.1 (continued)

Wind farm name	Years consulted	Biogeographic region	Number of wind turbines	Produced power (MW)
Meroicinha II	2006–2007	Subsector Miniense	6	15
Mértola	2015	Subsector Baixo-Alentejano-Monchiquense	19	43.7
Mosqueiros (I; II)	2009–2010	Sector Estrelense	14	28
Mougueiras e Bravo	2008–2009	Subsector Beirense Litoral	12	24
Pampilhosa da Serra	2009–2010	Subsector Beirense Litoral	38	114
Pedras Lavradas, Balocas e Sr.ª das Necessidades	2010–2011	Sector Estrelense	28	56
Picos-Vale do Chão	2010	Subsector Beirense Litoral	12	24.6
Pinhal Interior (I; II)	2005–2008	Subsector Hurdano-Zezerense	63	144
Portela do Pereiro	2015	Subsector Oeste-estremenho	4	7.2
Prados	2010	Sector Estrelense	17	39.1
S. Macário I	2007–2009	Subsector Miniense	5	11.5
São Pedro	2006–2008	Subsector Miniense	5	10
Serra Alta	2009–2011	Sector Lusitano-Duriense	1	2
Serra da Nave	2009	Subsector Miniense	19	38
Serra de Bornes	2010	Sector LUSITANO-DURIENSE	24	60
Serra de Candeeiros	2005–2008	Subsector Oeste-estremenho	37	111
Serra de Chavães e Sendim	2012–2013	Subsector Miniense	35	70
Serra do Barroso III	2010–2011	Subsector Geresiano-Queixense	8	16
Serra do Mú	2009	Subsector Baixo-Alentejano-Monchiquense	14	28.3
Serra do Sicó (I; II)	2009–2011	Subsector Oeste-estremenho	10	20
Sobrado	2010–2011	Subsector Miniense	4	8

(continued)

Table 1.1 (continued)

Wind farm name	Years consulted	Biogeographic region	Number of wind turbines	Produced power (MW)
Terra Fria	2008–2010	Subsector Geresiano-Queixense	52	104
Vale Estrela	2015	Sector LUSITANO-DURIENSE	11	25.3
Videira	2006–2008	Subsector Oeste-estremenho	3	6
Vila Nova II	2010–2012	Subsector Beirense Litoral	12	24

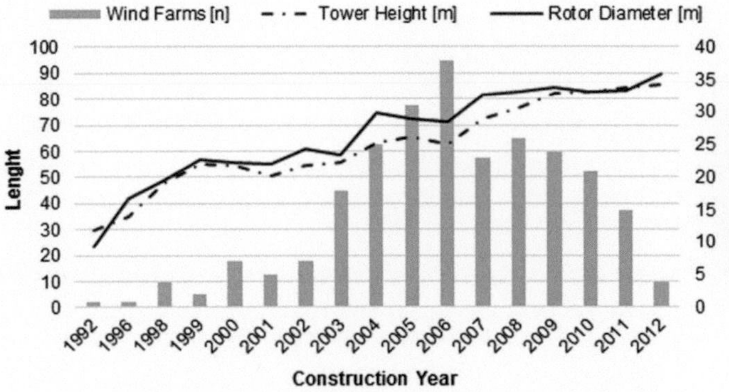

Fig. 1.3 Evolution of the wind turbine characteristics through time in relation to the number of wind farms constructed in Portugal (data from e2p 2015)

susceptible to negative impacts caused by wind farms than those found within SCIs. They are therefore subjected to a different assessment and evaluation during the environmental impact assessment process (see Santos et al., Chap. 3).

Wind farms in Portugal are considered small facilities, with an average of 10 wind turbines (min: 1; max: 57 turbines), with a generating capacity of 18 MW per facility (min: 0.02; max: 114 MW). Some facilities started in the 1990s with several small wind turbines, but most were constructed in the 2000s, with small- to medium-sized wind turbines (Fig. 1.3). Nowadays, most of the older wind farms have undergone repowering—lessening the number of installed wind turbines, but increasing the tower dimensions. Wind turbines in Portugal have approximately 70 m of hub height (min: 30 m; max: 95 m) and 75 m of rotor diameter (min: 24 m; max: 100 m) (Fig. 1.3).

1.1.2 Wind Farm Impacts on Wildlife

Although wind energy has advantages in mitigating climate change, wind farms are also associated with negative impacts on wildlife, including flying vertebrates as well as terrestrial mammals. These groups are affected differently but two main types of impacts have been reported for wind farm operation: direct impacts—fatalities caused by barotraumas and/or collision with rotating wind turbine blades, with the turbine tower or with the associated power lines and infrastructures; and indirect impacts—which include habitat loss and/or fragmentation or behavioural effects, such as avoidance of turbine area, barrier effect, decline of breeding success (Gove et al. 2013; Dai et al. 2015; Schuster et al. 2015).

Indirect impacts, including both habitat and behaviour related impacts, are usually detected by a decrease in the utilisation of the impacted area in comparison with similar undisturbed areas (Pearce-Higgins et al. 2009). In the specific case of

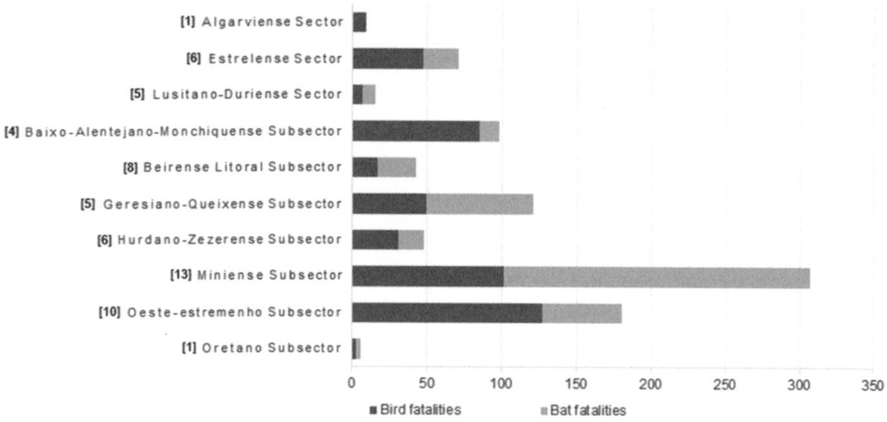

Fig. 1.4 Observed number of bird and bat fatalities (direct fatalities) per biogeographic region at the analysed wind farms in Portugal. The number of wind farms analysed for each region is presented [in brackets]

terrestrial mammals, indirect impacts are the most common reason for concern regarding wind farms, though an increase in expected mortality is also highlighted by some authors (Lovich and Ennen 2013). Such impacts may include habitat loss due to destruction and/or decreased quality, habitat fragmentation causing interruption in gene flow, and predator attraction, among others. In Portugal, most terrestrial mammals are not expected to be significantly affected by wind farm placement and operation; however, there is one mammal species that raises red flags: Iberian Wolf (*Canis lupus*). This species has been protected by a Portuguese Law since 1988 that prohibits killing or capturing individuals, as well as disturbing important areas for the species. Additionally, European Community legislation (i.e. Council Directive 92/43/EEC) includes the Iberian Wolf as a priority species in Annex II (see Costa et al., Chap. 5).

Thus, wind farm operation has shown through time impacts especially in relation to flying vertebrates. Perhaps because indirect impacts are more difficult to quantify and determine with certainty, direct impacts, i.e. fatality due to collision with wind turbines, are of highest concern. The most commonly used indicator to determine the level of impact caused by wind energy in Portugal is the observation of bird and bat fatalities. This information is available in bird and bat monitoring reports of operational wind farms in Portugal, which document and measure the number of observed fatalities around turbines. A previous approach to this indicator was presented in 2012, considering data from a different set of wind farms monitored between 2003 and 2010 (Bernardino et al. 2012). Our objective with this analysis is to take this preliminary analysis, broaden it spatially and temporally and provide a general quantification of the impacts caused by wind farms in Portugal on bird and bat species over the last 10 years of operation, answering the following questions: (i) where and when are bird and bat species colliding with wind turbines; (ii) which species are being particularly affected?

1.2 Bird and Bat Fatalities in Portugal

A total of 894 fatalities, 477 bird fatalities and 417 bat fatalities have been observed between 2004 and 2015 at the 44 wind farms analysed, which represent 548 of the total 2367 (23%) turbines installed in Portugal (Table 1.1). On average, each project conducted 2 years of operational phase monitoring (min. 1; max. 7). The summary of the dataset used for this analysis is presented in Fig. 1.4, where the number of bird and bat fatalities documented per biogeographic region in Portugal is shown. The fatality values presented are uncorrected for bias sources (e.g., carcass removal or searcher efficiency), since the specialists in Portugal tend to use different methodologies to calculate bias and estimate fatality (refer to Marques et al., Chap. 7). To present a comparable evaluation through wind farms, fatalities analysed in the next subchapters were weighted by the sampling effort (number of search visits and wind turbines searched for the total time length of the monitoring programme).

1.2.1 Biogeographic and Phenological Analysis of Bat Fatalities

Average bat fatalities weighted by sampling effort per biogeographic region are shown in Fig. 1.5. Most of the bat fatalities were found in four regions: the *Lusitano-Duriense* Subsector (0.08 bat fatalities/turbine/search); the *Geresiano-Queixense* Subsector and the *Beirense Litoral* Subsector (0.07 bat fatalities/turbine/search); and the *Hurdano-Zezerense* Subsector (0.06 bat fatalities/turbine/search). These are all located in the northern and northern-central part of Portugal.

Fatalities caused by wind farms may be related to the life cycle of the species. Three peaks of bat fatalities through the year are shown in Fig. 1.6: the highest peak of activity, between the second half of August and the second half of September, represents almost half of the fatalities found through the year (44%); the second peak of fatality is recorded in the second half of May (6%); while the third peak happens in the first half of April (3%). Relating these fatality events with the life cycle of the bat,[1] the highest peak matches the juvenile dispersion and mating season, while the first two peaks match the breeding season and the beginning of active night feeding, which is related to the rise of air temperature that consequently boosts insect populations and food availability for bats (Park et al. 2000; Celuch

[1]In Portugal, bat hibernation period is considered to start in mid-December and to last until late February; and breeding period starts from mid-April and goes as far as late July, depending on species (per ICNF 2014b). For the analysis presented in this chapter, the following periods were considered: hibernation: mid-December to end of February; breeding: March to June; dispersion: July to August; mating: September to mid-December.

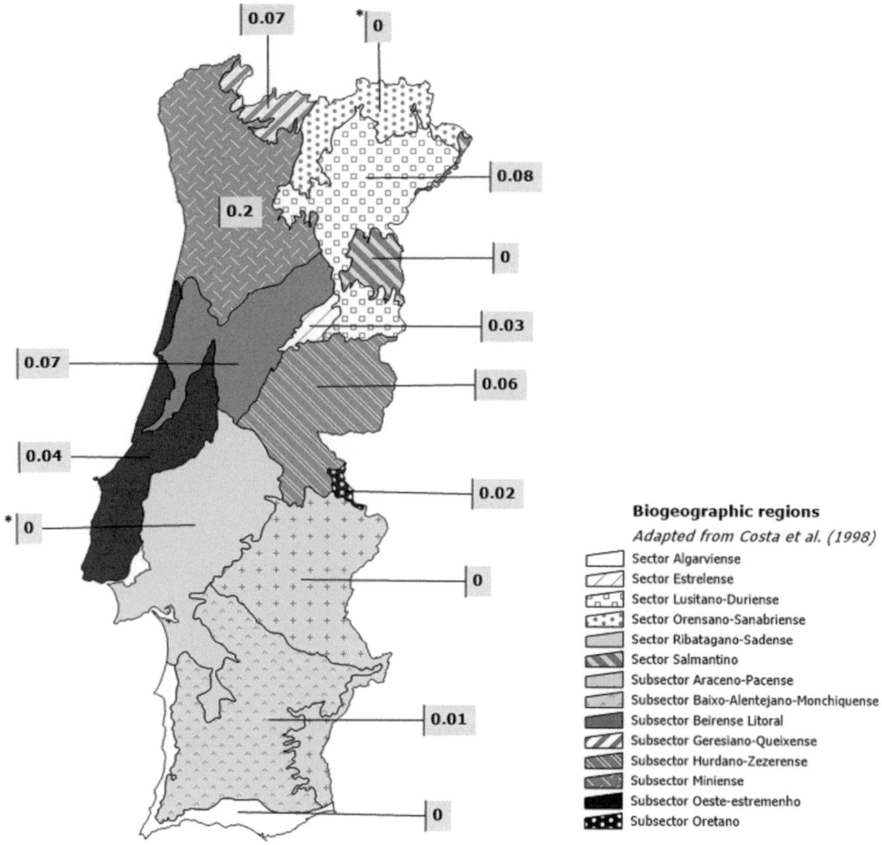

Fig. 1.5 Bat weighted fatality rates (number of fatalities/wind turbine/search) distribution in relation to biogeographic regions for the analysed wind farm projects in Portugal. *Asterisk symbol* (*) indicates the biogeographic regions where no wind farms were sampled for this analysis

and Kanuch 2005; Rodrigues and Palmeirim 2008). By the end of July, most bat species finish their breeding season and a large number of juvenile bats are active; while, in September, the mating season begins. This pattern of higher bat fatality in late summer and autumn has been observed all over the world (Dai et al. 2015; Schuster et al. 2015).

Bat activity is thus influenced by temperature and precipitation (Santos et al. 2013; Amorim et al. 2012), which are related to the biogeographic regions: Atlantic climate lowers temperatures and increases precipitation in the North, and Mediterranean climate increases temperatures and lowers precipitation in the South; a combination of both climates occur in the Centre. Figure 1.7 shows that this

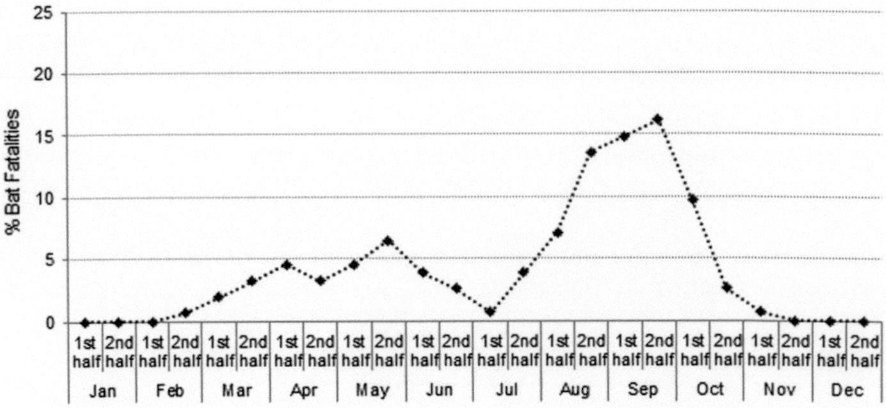

Fig. 1.6 Temporal distribution (Percentage of fatalities found throughout the year) of bat fatalities in each month of the year at the analysed wind farms in Portugal

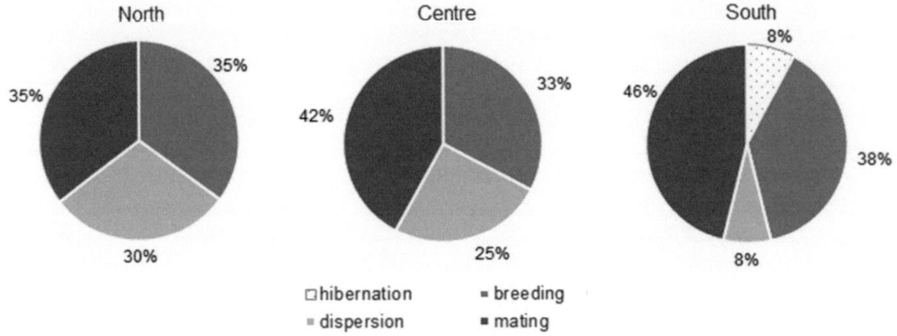

Fig. 1.7 Distribution of bat fatalities throughout Portuguese territory in relation to bat life cycle

gradient correlates with bat fatalities, since no fatalities occurred in the hibernation period in the North and Centre, since bats are likely to be inactive with the relative lower temperatures experienced in this part of the country. On the other hand, in the South, a considerable portion (8%) of bat fatalities were found during hibernation months (late February), which can be explained by a shorter bat hibernation period in the South when compared with the rest of the country. It is also noteworthy that both in the South and Centre most fatalities occurred during the mating season, suggesting that bats are more active in this period of their life cycle and are also more susceptible to colliding with rotating wind turbine blades.

1.2.2 Biogeographic and Phenological Analysis of Bird Fatalities

Bird fatalities weighted by sampling effort per biogeographic region are shown in Fig. 1.8. The spatial distribution differs slightly since most of bird fatalities were found in two regions: the *Miniense* Subsector (0.36 bird fatalities/turbine/search), and the *Estrelense* Sector (0.11 bird fatalities/turbine/search).

Fatalities distribution is also higher in late August and late September (28%), matching the theoretically highest bird abundance (adult and juvenile birds) and the

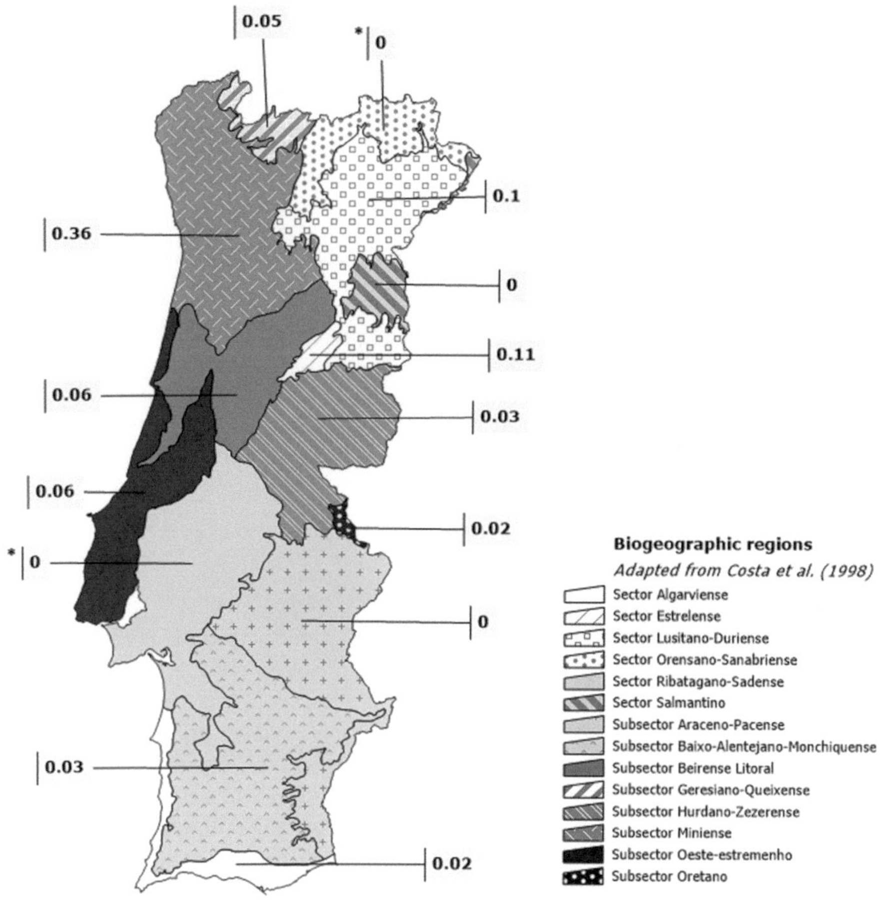

Fig. 1.8 Bird weighted fatality rates (number of fatalities/wind turbine/search) distribution in relation to biogeographic regions for the analysed wind farm projects in Portugal. *Asterisk symbol* (*) indicates the biogeographic regions where no wind farms were sampled for this analysis

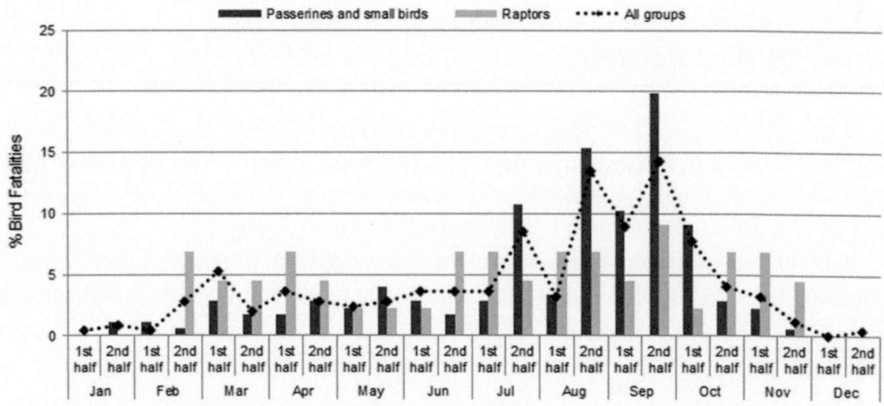

Fig. 1.9 Temporal distribution (percentage of fatalities found throughout the year) of bird fatalities in each month of the year at the analysed wind farms in Portugal

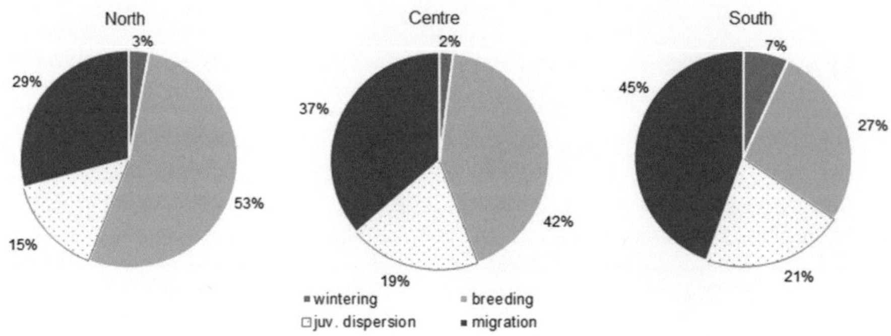

Fig. 1.10 Distribution of bird fatalities throughout Portuguese territory in relation to bird phenological seasons

migration season that occurs between late July and early December in Portugal[2] (Fig. 1.9). This trend is especially obvious in the South and Centre (Fig. 1.10), which may be related to the migration route of the southwestern region of Portugal (Sagres Peninsula). This is particularly important for juvenile and immature birds wintering in Africa, which travel to southern Europe to reach the preferential migration route through Strait of Gibraltar. A second peak of fatality is observed in late July, representing approximately 8% of all bird fatalities throughout the territory (Fig. 1.9), which may relate to the high abundance of juveniles and their dispersion or early migration movements. The third and final fatality peak is

[2]Bird phenological periods considered for the analysis presented in this chapter are the following: wintering: December to February; breeding: March to June; juvenile dispersion: July to August; migration: September to November.

observed in early March, when the breeding migrants arrive, especially in the northern regions of Portugal.

The Portuguese bird fatalities reflect the conclusions of recent studies that suggest that bird fatalities differ between groups (Fig. 1.9), i.e. raptors are more susceptible to collisions than passerines, and, in different timeframes, raptors suffer a higher rate of fatalities in spring while passerines are more affected during juvenile dispersion and migration (Wang et al. 2015). As shown in Fig. 1.9, raptor species fatalities were higher than passerines and small bird species between late February and late April, while the latter group fatalities were superior between late July and early October, including juvenile dispersion and migration period for many species.

1.2.3 Bat Fatalities Specific Composition

Bat species with the highest absolute fatalities and fatality rates are the *Pipistrellus* spp. and *Nyctalus* spp. genus, namely the Common pipistrelle (*Pipistrellus pipistrellus*) and the Lesser noctule (*Nyctalus leisleri*). Though there are no studies that quantify bat species abundance in mainland Portugal, the Common pipistrelle is classified as Least Concern in Portugal (Cabral et al. 2006), indicating that the species is quite abundant. On the other hand, the Lesser noctule is classified as Data Deficient, as there are no population studies on this species both in Portugal and in other European regions (Cabral et al. 2006). The species with the highest number of absolute as well as weighted fatalities was the Common pipistrelle (*Pipistrellus pipistrellus*; 133 fatalities and an average of 0.0030 fatalities/turbine/search) (Table 1.2). In fact, fatalities caused by wind farms are acknowledged to be one of the main causes of mortality for the species (Rainho et al. 2013; ICNF 2014a). Morphological separation between this species and the Pygmy pipistrelle is very difficult, since both species were considered as one in Portugal until 2002 (Salgueiro et al. 2002 in Rainho et al. 2013). For this reason, several clusters of species for this genus are presented, since it was not possible to morphologically distinguish between them, especially if they were in a decomposed state.

Pipistrellus spp. species are generally abundant and have a widespread distribution throughout the Portuguese territory (excluding the islands) (Rainho et al. 2013). Species of *Pipistrellus* spp. are easily observed within urban environments while feeding around street lights (Rydel and Racey 1995; Rainho et al. 2013), and as such their abundance, added to their tolerance of human infrastructures, is likely to contribute to a lack of avoidance of wind turbines, increasing the probability of collision with turbine blades.

Nyctallus spp. on the other hand is composed of species that are mostly associated with forest habitats, and have a higher avoidance of urban environments (Marques and Rainho 2006; Waters et al. 1999). These species usually fly at higher altitudes (Gloor et al. 1995; Palmeirim et al. 1999), which is a probable cause of the number of fatalities found, since they are more likely to collide with the rotating wind turbine blades. The association of these species to forested areas at high

Table 1.2 Top bat species absolute fatalities and weighted fatality rates (number of fatalities/wind turbine/search) at the analysed wind farms in Portugal between 2005 and 2015

Scientific name	Common Name	Weighted fatality		Absolute fatality		RL
		Number of fatalities	Rank	Total number of fatalities	Rank	
Pipistrellus pipistrellus	Common pipistrelle	0.0030	1	133	1	LC
Nyctalus leisleri	Lesser noctule	0.0027	2	101	2	DD
Hypsugo savii	Savi's pipistrelle	0.0021	3	22	5	DD
Pipistrellus sp.	–	0.0008	4	29	3	–
Pipistrellus pipistrellus/pygmaeus	–	0.0007	5	16	7	–
Pipistrellus kuhlii	Kuhl's pipistrelle	0.0007	6	26	4	LC
Nyctalus sp.	–	0.0006	7	9	8	–
Pipistrellus pipistrellus/pygmaeus/khuli	–	0.0005	8	3	11	–
Pipistrellus pygmaeus	Pygmy pipistrelle	0.0003	9	18	6	LC
Eptesicus serotinus	Serotine	0.0002	10	5	10	LC
Myotis daubentoni	Daubenton's bat	0.0001	11	2	15	LC
Eptesicus sp.	–	0.0001	12	6	9	–

RL (Portuguese Red List of Vertebrates, Cabral et al. 2006): *DD* Data Deficient; *LC* Least Concern; *VU* Vulnerable; *NA* Not Applicable

Table 1.3 Top bird species absolute fatality and weighted fatality rates (number of fatalities/wind turbine/search effort) at the analysed wind farms in Portugal between 2005 and 2015

Scientific name	Common name	Weighted fatality		Absolute fatality		RL	Phenology
		Number of fatalities	Rank	Total number of fatalities	Rank		
Delichon urbicum	Northern house-martin	0.00439	1	158	1	LC	BM
Apus apus	Common swift	0.00203	2	18	6	LC	BM
Alauda arvensis	Eurasian skylark	0.00108	3	21	4	LC	R
Lullula arborea	Woodlark	0.00087	4	25	3	LC	R
Sylvia undata	Dartford warbler	0.00077	5	7	10	LC	R
Alectoris rufa	Red-legged partridge	0.00049	6	19	5	LC	R
Buteo buteo	Eurasian buzzard	0.00032	7	13	7	LC	R
Gyps fulvus	Griffon vulture	0.00032	8	12	8	NT	R
Falco tinnunculus	Common kestrel	0.00024	9	39	2	LC	R
Ficedula hypoleuca	European pied flycatcher	0.00015	11	8	9	NA	NBM
Phylloscopus collybita + ibericus	Common chiffchaff + Iberian chiffchaff	0.00014	12	7	10	LC	WM/ BM
Circus pygargus	Montagu's harrier	0.00005	20	7	10	EN	BM

RL (Portuguese Red List of Vertebrates, Cabral et al. 2006): *DD* Data Deficient; *LC* Least Concern; *NT* Near Threatened; *EN* Endangered; *CR* Critically Endangered; Phenology (Equipa Atlas2008): *R* Resident; *BM* Breeding Migrant; *WM* Wintering Migrant; *NBM* Non-Breeding Migrant

altitudes, with a high abundance of water sources, may be a determining risk factor, since most wind farms in Portugal are in environments with these characteristics.

It is of note that no species with an unfavourable conservation status is included in the list of species more frequently found dead at the Portuguese wind farms monitored. However, it is also noteworthy that five of the listed species have insufficient available data regarding their populations, hence their conservation status was not evaluated (Cabral et al. 2006). One of these species, the Lesser noctule (*Nyctalus leisleri*), is included as the second species with the highest absolute as well as weighted fatality found at Portuguese wind farms (101 fatalities), with an average of 0.0027 fatalities/turbine/search (Table 1.2).

1.2.4 Bird Fatalities Specific Composition

Bird fatalities are mostly composed of two groups, accounting for two thirds of all bird fatalities found: 'Swallows, martins and swifts' and 'Larks' (Table 1.3). The first group are breeding migrants that primarily occur in Portugal between March and September, but may arrive as early as January (Equipa Atlas 2008). Despite being present only a few months of the year, these species reach higher fatality rates than resident species. Swallows, martins and swifts share behavioural traits related to their foraging and flight habits: these are aerial insectivorous feeders that form large aerial flocks (SEO 2008). As they hunt for small aerial insects that are attracted to wind turbines (Long et al. 2011), they enter the collision risk zones, which explains the high collision rates found.

In relation to larks, most of these species are resident in Portugal, and they are mostly present in the mountainous areas throughout the year (Equipa Atlas 2008). The most affected lark species are the Eurasian skylark (*Alauda arvensis*) and the Woodlark (*Lullula arborea*) (Table 1.3). Like swallows, martins and swifts, some lark species also tend to form flocks, especially in the post-breeding season, for migration (Hyman 2006). Larks display during breeding season while they perform aerial flights at considerable altitudes, sometimes hovering in the air several dozens of meters above ground (Camfield 2004). The Woodlark has similar habits, hence enhancing the importance that this trait may have to explain the high fatality rates of this group (SEO 2008). On the other hand, considering that most wind farms in Portugal are located within mountainous areas, where this group is more abundant, they also have a higher collision risk due to an increased exposure to risk. A recent study in Portuguese wind farms have shown through simulation that wind turbines are likely to cause fatalities of up to 4% on Eurasian skylark breeding populations in 2026, due to the increasing number of wind turbines in the habitat areas for the species (Bastos et al. 2015).

Besides passerines and small birds, birds of prey fatalities are of special concern. This is the case for the ninth species in terms of highest weighted fatality (average 0.00024 fatalities/turbine/search), the Common kestrel (*Falco tinnunculus*) with 39 fatalities (Table 1.3). This species hovers at rotor swept area to hunt for prey, which increases the collision risk and explains the high fatality rate. Though this species has a relatively lower number of fatalities and weighted fatality, other top-ranked fatalities species in Portugal include the Common buzzard (*Buteo buteo*; ranked in 7th for weighted fatality) and the Griffon vulture (*Gyps fulvus*; ranked in 8th for weighted fatality) (Table 1.3). However, higher longevity and lower breeding rates of these species suggest a higher magnitude of impacts. As per the Portuguese Red List of Vertebrates (Cabral et al. 2006), the only two species with an unfavourable conservation status included in the list are the Montagu's harrier, an Endangered species with seven fatalities (ranked in 10th for number of total fatalities); and the Griffon vulture, a Near Threatened species with eight fatalities (ranked 8th for number of total fatalities) (Table 1.3).

1.3 Final Notes

The general characterisation of the wind farm industry and its impacts on wildlife in continental Portugal reported in this chapter reveals that the highest fatality rates are from common, abundant bird and bat species with a favourable conservation status. However, some facts raise concern:

- The lack of knowledge on the population size of several bat species that prevent the classification of a conservation status also hinders the assessment of the real impact of the wind farm fatalities on their population viability;
- As bat fatalities were found to happen during hibernation period in southern Portugal, it is possible that impacts are occurring throughout the year (especially during warmer winter years), instead of only between March and October, when bats are expected to be active. Thus, it could be important to review the hibernation periods for bats, especially at wind farms located in southern Portugal, possibly increasing the duration of the monitoring effort of the bat monitoring guidelines.
- Few wind farm fatalities on long-lived bird species with unfavourable conservation status (e.g., Montagu's harrier, Griffon vulture) may have important effects on their population viability (Carrete et al. 2009), which is not being assessed;
- The cumulative impact of wind farm fatalities on bird species ranked as highly vulnerable to collision in Portugal as well in Spain (Lucas et al. 2009, 2012a, b) upgrades the impact to the Iberian populations level, since several of this species are long-distance migrants or dispersive birds (e.g., kites and vultures). The lack of a wide and systematic monitoring leads to an absence of impact evaluation at this broader scale.
- The same problems regarding cumulative impacts affects bat populations, since several bat species occurring in Portugal may have local and regional movements, such as the Schreiber's bat or the noctule bats (Ramos Pereira et al. 2009; Rodrigues and Palmeirim 2008; Voigt et al. 2015). Though the Schreiber's bat

is not highly affected by fatalities at wind farms, noctule bats are included amongst the species with the highest fatalities detected. Since little is known about the species movements in Portugal and the Mediterranean, nor its population size, no impact evaluation has been carried out regarding impacts on such species on the broader scale.

Wind farms have been causing negative impacts on bird and bat populations since they first started to be installed in Portugal in the 1990s. Since the early stages of wind farm development and impact assessment, a major question arose regarding the prediction of impacts and suggestions for the best locations for wind farms: what parameters help to explain and model bird and bat fatality distribution and occurrence in Portugal? This preliminary analysis of impact characteristics provides the first clues to the answer, but also to the definition of mitigation and compensation measures. Bird and bat fatalities may be site- and species-specific: at the same location, different groups of turbines may show different risk levels (Dai et al. 2015; Schuster et al. 2015; Wang and Wang 2015), and the analysis conducted is in line with these findings. However data with enough detail to provide further insight into these micro-scale influences are difficult to obtain and analyse, such as behavioural data, and micro-scale landscape and meteorological data. As an example, landscape modification, turbine height, and severe weather conditions have been said to influence bat fatality occurrence in other locations (Wang and Wang 2015). Future research should improve methods to collect and analyse this information, and should focus on studying the relation of bird and bat fatalities with landscape features (particularly habitat changes), time of the year, environmental conditions and characteristics of wind farms and/or wind turbines. The influence of all these factors on bird and bat fatalities remain unanswered, both in Portugal and worldwide (Wang et al. 2015).

References

Amorim, F., Rebelo, H., & Rodrigues, L. (2012). Factors influencing bat activity and mortality at a wind farm in the Mediterranean region. *Acta Chiropterologica, 14,* 439–457.

Bastos, R., Pinhanços, A., Santos, M., Fernandes, R. F., Vicente, J. R., Morinha, F., et al. (2015). Evaluating the regional cumulative impact of wind farms on birds: How can spatially explicit dynamic modelling improve impact assessments and monitoring? *Journal of Applied Ecology, 53,* 1330–1340.

Bernardino, J., Zina, H., Costa, H., Fonseca, C., Pereira, M. J., & Mascarenhas, M. (2012). Bird and bat mortality at Portuguese wind farms. Poster presented at the 32nd Annual Conference of the International Association for Impact Assessment, Centro de Congressos da Alfândega, Porto, Portugal, 27 May–1 June 2012.

Cabral, M. J. (coord.), Almeida, J., Almeida, P. R., Dellinger, T., Ferrand de Almeida, N., Oliveira, M. E., et al. (Eds.) (2006). Livro Vermelho dos Vertebrados de Portugal 2ª ed. Instituto da Conservação da Natureza/Assírio & Alvim. Lisbon [In Portuguese].

Camfield, A. (2004). *Alaudidae. Animal Diversity Web.* Accessed February 20, 2017. Retrieved from http://animaldiversity.org/accounts/Alaudidae/

Carrete, M., Sánchez-Zapata, J. A., Benítez, J. R., Lobón, M., & Donázar, J. A. (2009). Large scale risk-assessment of wind-farms on population viability of a globally endangered long-lived raptor. *Biological Conservation, 142,* 2954–2961.

Celuch, M., & Kanuch, P. (2005). Winter activity and roosts of the noctule (Nyctalus noctula) in an urban area (Central Slovakia). *Lynx (Praha), 36,* 39–45.

Costa, J. C., Aguiar, C., Capelo, J. H., Lousã, M., & Neto, C. (1998) Biogeografia de Portugal Continental. Quercetea: 5–56 [In Portuguese].

Dai, K., Bergot, A., Liang, C., Xiang, W.-N., & Huang, Z. (2015). Environmental issues associated with wind energy—A review. *Renewable Energy, 75,* 911–921.

Equipa Atlas. (2008). Atlas das Aves nidificantes em Portugal (1999–2005). Instituto da Conservação da Natureza e da Biodiversidade, Sociedade Portuguesa para o Estudo das Aves, Parque Natural da Madeira e Secretaria Regional do Ambiente e do Mar. Assírio & Alvim. Lisbon. [In Portuguese].

European Parliament (EU). (2009). Directive 2009/28/EC of the European Parliament and of the Council of 23 April 2009 on the promotion of the use of energy from renewable sources and amending and subsequently repealing Directives 2001/77/EC and 2003/30/EC (Text with EEA relevance).

E2P—Energias Endógenas de Portugal. (2015). Wind Farms in Portugal. INEGI—Institute of Mechanical Engineering and Industrial Management & APREN—Portuguese Renewable Energy Association.

Gloor, S., Stutz, H.-P. B., & Ziswiller, V. (1995). Nutritional habits of the noctule bat Nyctalus noctula. *Myotis, 32–33,* 231–242.

Gove, B., Langston, R. H. W., McCluskie, A., Pullan, J. D., & Scrase, I. (2013). *Wind Farms and Birds: An updated analysis of the effects of wind farms on birds, and best practice guidance on integrated planning and impact assessment.* RSPB/BirdLife in the UK: Report prepared by BirdLife International on behalf of the Bern Convention.

Hyman, D. (2006). *Alauda arvensis. Animal Diversity Web.* Accessed February 20, 2017. Retrieved from http://animaldiversity.org/accounts/Alauda_arvensis/

International Renewable Energy Agency (IRENA). (2016). *Renewable capacity highlights.* Retrieved from http://resourceirena.irena.org/gateway/

ICNF. (2014a). Agreement on the Conservation of Populations of European Bats. Report on implementation of the Agreement in Portugal. 2014/7 MoP.

ICNF. (2014b). Análise dos dados do Programa de Monitorização de Abrigos Subterrâneos de Importância Nacional de Morcegos (1988–2012). Instituto da Conservação da Natureza e das Florestas [In Portuguese].

ICNF. (2017). *Áreas Protegidas, Rede Natura e Sítios Ramsar - Portugal continental.* http://www.icnf.pt/portal/naturaclas/cart/ap-rn-ramsar-pt. Accessed February 20, 2017 [In Portuguese].

Long, C. V., Flint, J. A., & Lepper, P. A. (2011). Insect attraction to wind turbines: does colour play a role? *European Journal of Wildlife Research, 57,* 323–331.

Lovich, J. E., & Ennen, J. R. (2013). Assessing the state of knowledge of utility-scale wind energy development and operation on non-volant terrestrial and marine wildlife. *Applied Energy, 103,* 52–60.

Lucas, M., Janss, G. F. E., & Ferrer, M. (Eds.). (2009). *Aves y Parques Eólicos.* Quercus, Madrid: Valoración del Riesgo y Atenuantes. ISBN 978-84-87610-19-6.

Lucas, M., Ferrer, M., Bechard, M. J., & Muñoz, A. R. (2012a). Griffon vulture mortality at wind farms in southern Spain: Distribution of fatalities and active mitigation measures. *Biological Conservation, 147,* 184–189.

Lucas, M., Ferrer, M., & Janss, G. F. E. (2012b). Using wind tunnels to predict bird mortality in wind farms: The case of griffon vultures. *PLoS ONE, 7,* e48092.

Marques, J. T., & Rainho, A. (2006). GAPS – Gestão Ativa e Participada do Sítio de Monfurado. CMMN & ICN, Montemor-o-Novo [In Portuguese].

Marques, A. T., Batalha, H., Rodrigues, S., Costa, H., Pereira, M. J. R., Fonseca, C., et al. (2014). Understanding bird collisions at wind farms: An updated review on the causes and possible mitigation strategies. *Biological Conservation, 179,* 40–52.

Palmeirim, J. M., Rodrigues, L., Rainho, A., & Ramos, M. J. (1999). Chiroptera, in Guia dos Mamíferos Terrestres de Portugal Continental, Açores e Madeira, Mathias ML (Ed.) Instituto da Conservação da Natureza, Centro de Biologia Ambiental da Universidade de Lisboa, Lisboa [In Portuguese].

Park, K. J., Jones, G., & Ransome, R. D. (2000). Torpor, arousal and activity of hibernating Greater Horseshoe Bats (Rhinolophus ferrumequinum). *Functional Ecology, 14,* 580–588.

Pearce-Higgins, J. W., Stephen, L., Langston, R. H. W., Bainbridge, I. P., & Bullman, R. (2009). The distribution of breeding birds around upland wind farms. *Journal of Applied Ecology, 46,* 1323–1331.

Presidência do Conselho de Ministros (PCM). (2013). Resolução do Conselho de Ministros n.º 20/2013: Plano Nacional de Ação para as Energias Renováveis (Estratégia para as Energias Renováveis - PNAER 2020). Diário da República, 1st series—Nr 70: 2022–2091 [In Portuguese].

Rainho, A,, Alves, P., Amorin, F., Marques, J. T. (Coord.). (2013). Atlas dos morcegos de Portugal Continental. Instituto da Conservação da Natureza e das Florestas. Lisbon. 76 pp + Appendixes [In Portuguese].

Ramos Pereira, M. J., Salgueiro, P., Rodrigues, L., Coelho, M. M., & Palmeirim, J. M. (2009). Population structure of a cave-dwelling bat, Miniopterus schreibersii: Does It reflect history and social organization? *Journal of Heredity, 100,* 533–544.

Rodrigues, L., & Palmeirim, J. M. (2008). Migratory behaviour of the Schreiber's bat: when, where and why do cave bats migrate in a Mediterranean region? *Journal of Zoology, 274,* 116–125.

Rydell, J., & Racey, P. A. (1995). Street lamps and the feeding ecology of insectivorous bats. *Ecology, Evolution and Behaviour of Bats, 67,* 291–307.

Santos, H., Rodrigues, L., Jones, G., & Rebelo, H. (2013). Using species distribution modelling to predict bat fatality risk at wind farms. *Biological Conservation, 157,* 178–186.

Schuster, E., Bulling, L., & Köppel, J. (2015). Consolidating the state of knowledge: A synoptical review of wind energy's wildlife effects. *Environmental Management, 56,* 300–331.

SEO/BirdLife. (2008). La enciclopédia de las aves de España. Fundacion BBVA, SEO/BirdLife. ISBN-13: 978-84-936441-1-6 [In Spanish].

Frankfurt School-UNEP Centre/BNEF. (2016). *Global trends in renewable energy investment 2016*. Retrieved from http://www.fs-unep-centre.org

Voigt, C. C., Lehnert, L. S., Petersons, G., Adorf, F., & Bach, L. (2015). Wildlife and renewable energy: German politics cross migratory bats. *European Journal of Wildlife Research, 61,* 213–219.

Wang, S., & Wang, S. (2015). Impacts of wind energy on environment: A review. *Renewable and Sustainable Energy Reviews, 49,* 437–443.

Wang, S., Wang, S., & Smith, P. (2015). Ecological impacts of wind farms on birds: Questions, hypotheses, and research needs. *Renewable and Sustainable Energy Reviews, 44,* 599–607.

Waters, D., Jones, G., & Furlong, M. (1999). Foraging ecology of Leisler's bat (Nyctalus leisleri) at two sites in southern Britain. *Journal of Zoology, 249,* 173–180.

Chapter 2
Impacts of On-shore Wind Farms in Wildlife Communities: Direct Fatalities and Indirect Impacts (Behavioural and Habitat Effects)

Pedro Pereira, Nuno Salgueiro and Sílvia Mesquita

2.1 Introduction

Wind energy is considered an infinite resource with plenty of environmental advantages. It is a free source of clean energy because, in contrast to other types of energy production, no GHG or pollutant particles are emitted into the atmosphere during its production. It has an important role to reduce the global emissions of GHG to the atmosphere in the near-term (until 2020) but also in the long-term (2050) (Arvizu et al. 2011). Additionally, no discharge to water or soil or solid waste is produced (Ledec et al. 2011).

The European wind power industry has grown rapidly to meet EU targets of sourcing 20% of energy from renewable sources by 2020. The energy strategy defined by the Portuguese Government in 2013 set the mark even further, aiming for 31% in 2020 (DGEG 2016a). In fact, increasing renewable energy production has become a major goal for Portuguese government for the past 15 years to reduce external dependence on energy supply. In 2015, renewable production in Portugal achieved 52% of national electrical consumption. Comparing the increase of wind energy production in recent years, in 2006, around 5% of electricity consumed was produced by wind farms; while in 2015, this production represented 23% (DGEG 2016a).

Although undeniably a green resource, wind energy entails adverse effects, scientifically or non-scientifically addressed, on humans, landscape and wildlife.

P. Pereira · N. Salgueiro · S. Mesquita (✉)
Bioinsight, Odivelas, Portugal
e-mail: silvia.m@bioinsight.pt

P. Pereira
e-mail: pedro.p@bioinsight.pt

N. Salgueiro
e-mail: nuno.s@bioinsight.pt

M. Mascarenhas et al. (eds.), *Biodiversity and Wind Farms in Portugal*,
https://doi.org/10.1007/978-3-319-60351-3_2

One of the well-studied affected groups is wildlife, with scientists focusing mostly on bats, migratory birds, breeding and resting birds, raptors and marine mammals (Schuster et al. 2015). Wind energy projects are still expanding worldwide, and scientists, environmental technicians and entities are still working into find answers for negative effects and causes and to delineate mitigation measures.

Although there are few published studies in Portugal, this chapter points out the known impacts of on-shore wind energy that affect birds, bats and terrestrial mammals, considering the portuguese reality. A little contextualization about impact evaluation is made for framing purposes.

2.2 Evaluating Impacts

The International Association for Impact Assessment defines Environmental Impact Assessment (EIA) as the process of identifying, predicting, evaluating and mitigating the biophysical, social, and other relevant effects of development proposals prior to major decisions and commitments (IAIA 1999). The immediate purpose of EIA is to supply decision-makers with an indication of the likely environmental consequences of their actions, with the aim of ensuring that development only proceeds in an acceptable manner (Jay 2007). EIA is well established worldwide and almost every nation has implemented its own environmental law process. International law and lending institution standards have also incorporated EIA as requirement and decision instrument. The use of EIA at different levels of decision-making is growing significantly, as is the range of decision-types for which it is now used (Morgan 2012).

Assessing potential impacts follows the logic of sustainable development, aiming to ensure the environment protection and contributing to improve the quality of human life. In wind farms, this evaluation aims to harmonize energy production with biodiversity conservation and management (Marques et al. 2014). Wind energy unintended side effects on wildlife have long been discussed and substantial research has evolved over the last decade (Schuster et al. 2015).

An accurate prediction of the potential impacts demands a comprehensive collection of detailed baseline data. High quality data is vital to understanding the potential impact and correctly informing decision makers, stakeholders and the public. By contrast, low quality information has been a major cause for litigation and economic loss (Chang et al. 2013). Monitoring programmes during the project construction and operation phases allow and confirm the occurrence of predicted impacts, and the eventual identification of additional impacts (Marques et al. 2014). Once identified, impacts can be reduced following a mitigation hierarchy: first avoid the impacts, then minimise and, as a last resort, compensate the residual impacts that could not be minimised (Rodrigues et al. 2015, Paula et al. 2015).

In many countries, national and regional entities provide guidelines and best practice recommendations for the EIA process (e.g., Atienza et al. 2011; Strickland et al. 2011), giving guidance on methodological options for field survey and on the identification and classification of potential impacts (Marques et al. 2014).

In 1987, the Portuguese government recognized the EIA as an instrument for environment policy and land management, but only in 1990 was it established by law (APA 2011).

2.3 On-shore Wind Farm Impacts

In the last three decades, the world has seen a global expansion of wind energy production. Besides energy global targets, accelerated development of technologies, continuous reduction of costs in transportation and assembly of turbines have fed this growth. At the same time, some issues have been identified, with special focus on some negative interactions between turbines and people or biodiversity, worrying countries, promoters, investigators, and all the other stakeholders in many parts of the world, especially in some developing countries and ecologically vulnerable regions (Bourillon 1999; Dai et al. 2015).

Regarding human health, the noise of wind turbines in particular has raised concerns, as well as the shadows produced by rotating blades. The adoption of prevention measures prior to wind farm construction (e.g., that they be a minimum distance from human settlements and observe noise guidelines) is expected to prevent any health hazard to human populations. Nevertheless, annoyance effects have been reported from populations living near wind turbines, but to date no scientific study has supported direct health hazards resulting from wind farm facilities (Ryberg et al. 2013; Knopper and Ollson 2011).

Concerning landscape, different negative aspects of wind farms may be pointed out by the different stakeholders (e.g., citizens, scientists, policy makers). Landscape is a complex concept that encompasses not only natural values but also sociocultural values, and these impacts should not be restricted merely to visual impacts. Recognizing this, several efforts have been made to integrate scientific values but also perceived values in landscape planning and analysis in the context of wind farm development, narrowing the gap between expert and non-expert opinion (Ryberg et al. 2013). In the European Union, the 'European Landscape Convention' was created, integrating the public perceptions and evaluations of landscape, and raising awareness for the public participation in the decisions that affect landscapes (COE 2000).

Impacts of wind energy on biodiversity are also known and well studied. They have been identified for a great range of situations since construction to operation phases of wind farms, affecting species ranging from plant to invertebrates, and from marine (off-shore wind farms) to terrestrial or flying vertebrates. The most affected are species whose populations are under stress and present unfavourable conservation status, species potentially more susceptible to the impacts of wind farms, which includes species with behavioural or eco-morphological traits that increase collision risk (e.g., open space foraging bats or migrants that fly at rotor sweep zone; raptors that spend long periods foraging at rotor sweep; and birds with low manoeuvrability flights); and regionally or locally rare species, or common species with declining populations (Marques et al. 2015).

A high number of surveys have been carried out within this field of research, focusing particularly on bats, birds and some terrestrial mammals in the case of on-shore wind-farms, plus marine mammals, fish and benthos in the case of off-shore wind-farms. These studies were aimed at understanding the mechanisms behind the negative effects and using this knowledge for the development of mitigation measures and planning tools. Hereafter, it is essential to resume what we have learnt so far and which doubts remain in order to allow for a sensitive wind energy development and to focus future research on lacks so far (Schuster et al. 2015).

Often, the impacts are only assessed independently, without a global analysis of another or other sources of impacts that collectively or, over more time, may have greater repercussions. A cumulative impact takes place when an individual effect is augmented due to effects that have occurred in the past, are presently occurring or that are predicted to happen in the future, independently of its source. These types of impacts may be classified individually as of low significance but during a particular time period, collectively, may be classified has significant (APA 2009; Canter 1996). Although cumulative impacts are increasingly included within EIAs, the quality of these assessments remains far from adequate (Piper 2001; Masden et al. 2009).

When analysing cumulative impacts of a wind farm, other existing or planned wind farms and associated power lines should be accounted for, as well as other type of infrastructures depending on their proximity and degree of effects or impacts. When determining the study area for cumulative impact assessment, a cartographic analysis is necessary, considering factors such as orography, road presence, other projects or land characteristics that might create discontinuity or barriers on the territory. Among the most common effects producing significant cumulative impacts are biodiversity and landscape. As to fauna, cumulative impacts are more sensitive in rising fatalities (direct impacts), disturbance and/or displacement, habitat loss and/or fragmentation. Concerning the Portuguese experience, the preferred location for onshore wind farms of mountain tops, with particular weather and geophysical conditions that allow less common flora to surge in restricted areas, may promote cumulative impacts on habitats and flora (APA 2009).

2.3.1 *Direct Impacts*

Direct fatalities represent the major threat of wind farms on flying vertebrate's communities. Many studies in North America and Europe have demonstrated over the past years that many bats and birds are killed at wind farms. In Portugal between 2003 and 2015, 1001 bat carcasses were found in a universe of 7704 found in Europe (EUROBATS 2016), while between, 2003 and 2010, 200 bird fatalities were recorded in Portugal (Bernardino et al. 2012). Global estimates of bird deaths from collisions with WT worldwide indicates 0.3 deaths per gigawatts/hour (GW/h)

(Sovacool 2009). The major cause of death of both birds and bats is traumatic injuries caused by collision against turbines (Rollins et al. 2012). Nevertheless, in the specific case of bats some theories suggest barotrauma as a cause (Baerwald et al. 2008), although with minor importance (Rollins et al. 2012).

Before understanding the cause of death, investigators' concern is to find reasons for the collisions. At this point, investigations are usually made ecological detachment.

2.3.1.1 Bat Collisions

Cryan and Barclay (2009) underlined that bat collisions can occur for different reasons, i.e. random collisions, coincidental collisions and collisions resulting from attraction. In the first case, collisions can be explained by high concentrations of individuals in a certain area increasing the likelihood for collisions. Thus, collisions emerge from chance events without an associated behavioural cause. Coincidental collisions have behavioural factors behind them, such as migration, mating or feeding, or even landscape features. Finally, a third group is classified as collisions resulting from attraction, once some elements of turbines can attract flying vertebrates and increase the collision probability (e.g., the turbine nacelle are used for roost or base for nests).

All three theoretical causes were sustained by a few practical cases, although a consensus was not always found. Random collisions, for example, were supported by the findings of Baerwald and Barclay (2011) in Canada who linked the increase of fatalities of silver-haired bats (*Lasionycteris noctivagans*) with increased activity. However, Piorkowski and O'Connell (2010) rejected the random collision hypothesis as they did not find any relation between fatalities and the high density of flying Brazilian free-tailed bats (*Tadarida brasiliensis*) in the USA. Thus, random collisions could be the cause of some fatalities, although not always with great significance, once some aspects as wind farm layouts and bat behaviours would take a most important role.

Coincidental collisions seem very logical. It is well known that, in nature, some behaviours increase the probability of some events. At the same time, it makes sense that settings from the environment will condition the activities and vulnerabilities of living beings. In this way, landscape features and wind farm layouts, as well as behaviours such as mating or feeding, could make bats more vulnerable to colliding with turbines. Piorkowski and O'Connell (2010) found that an eroded ravine, somehow serving as a conduit for the daily movements of many bats, was the best way to explain a hot spot of collision mortality in the USA.

Finally, regarding collisions resulting from attraction, some investigations have addressed the problem. Cryan et al. 2014 hypothesized a possible misperception that tree bats have of wind streams produced by wind turbines: they may respond to streams of air flowing downwind from trees at night while searching for roosts, conspecifics and nocturnal insect prey that could accumulate in such flows. This theory was tested in Portugal (Candeeiros wind farm) (Correia et al. 2014), through an acoustic study with Vestas turbines (model V90). Investigators did not find a source of any ultrasound signal that eventually could attract bats. However, bats hearing system is quite sensitive to the Doppler effect caused by the rotating blades. Although there are little attraction evidences, this hypothesis is still deserving of further investigation.

2.3.1.2 Bird Collisions

Bird collisions have been documented for a wide range of taxa and countries. Recent review works were published concerning this subject. Marques et al. (2014) grouped collisions into three main types: site-, wind farm- and/or species-specific factors. A brief explication of the three may be that collisions can be influenced for site-specific factors such as landscape features, food availability and weather; wind-farm factors are deeply connected with wind farms layout or turbine features; and species-specific factors are connected with specific physical and behavioural characteristics and how each species reacts.

Regarding site-specific factors, collisions may increase when turbine areas are within landscape features that are frequently used by birds (Marques et al. 2014; Schuster et al. 2015; Smith and Dwyer 2016), such as valleys or steep slopes. For example, in Portugal, the influence of topography and wind on habitat selection of common kestrels (*Falco tinnunculus*) were tested in Candeeiros and Chão Falcão I e II wind farms between 2008 and 2012. The tested variables play an important role for hunting kestrels since such birds chose to hunt mainly on wind-facing slopes with open habitats (Cordeiro et al. 2012), increasing sits vulnerability in a danger wind turbine situation. Such preference may explain the high incidence of fatalities of common kestrels at some Portuguese wind turbines.

Bird collisions can also be explained by wind farm factors, such as turbine characteristics. In some studies, fatalities increased with turbine height (de Lucas et al. 2008; Thelander et al. 2003) and rotor speed (Thelander et al. 2003). The turbine rotor diameter may also increase bird mortality, as it increases the area where birds are at risk (Loss et al. 2013). Wind farm factors have similarities with landscape factors, since both condition birds' activity in normal actions like hunting or matting, or simply movement.

Different behaviours may induce collision risk; for example. some body characteristics such as large birds are associated with less agility. Focusing on behaviours, Hull et al. (2013) concluded that species that forage on the ground have a lower probability of collision compared with species that forage in the air. For species with visual fields that may prohibit them from detecting turbines, the collision risk is bigger; this is also true for large and less agile species with weak-powered flight, which restricts its maneuverability. Vultures in the genus *Gyps* have both characteristics (De Lucas et al. 2008; Martin 2011; Martin et al. 2012), which makes them a genus of high vulnerability.

2.3.2 Indirect Impacts

Indirect impacts are often produced away from, or as a result of, a complex impact pathway. Such impacts may have consequences in the stress of individuals, affecting their fitness. Drewitt and Langston (2006), besides collision, identified displacement, barrier effects, habitat change and habitat loss as the main effects wind farms can have on birds, but in fact, all of these may also affect bats and some terrestrial mammals.

Disturbance and/or displacement may occur during both construction and operation phases and are associated with behaviour factors. It has been documented for several species from flying vertebrates (e.g. Kowallik and Borbach-Jaene 2001; Larsen and Madsen 2000; Leddy et al. 1999; Masden et al. 2009; Pearce-Higgins et al. 2009) to terrestrial mammals (e.g. Helldin et al. 2012; Álvares et al. 2017) that change their normal activity due to sensitivity to noise and machinery movement, tall structures, turbine noise, visual flicker and shadow effects. In Portugal, a study carried out through the analysis of 39 wind farm monitoring reports (Cordeiro et al. 2010) found this impact in six cases. In 10 of the 39, disturbance and/or displacement was not observed and in 11 it was not mentioned. The impact cases include birds and bats, as well as two cases of disturbance of the Iberian wolf (Cordeiro et al. 2010). Among terrestrial mammals, the Iberian wolf has a special position, not only for its unfavourable conservation status, but mostly because of its sensitivity to anthropogenic disturbance. In Portugal, there are about 1200 turbines in the wolf distribution area (Pimenta et al. 2005). The increase of wind farms in Portugal constitutes an issue for wolf conservation (Eggermann et al. 2011), since the species distribution area is located in mountain ridges, that simultaneously has great wind power potential (see more in Chap. 5).

Wind farm construction also induces land transformation during the whole lifetime of the wind farm, resulting in habitat loss and/or fragmentation. In Portugal, there are about 2711 on-shore turbines installed in the continental area (DGEG 2016b), which means a number near 56 Kw/Km^2, which is above that of countries such as Spain, Italy, France, the UK, Sweden and Ireland (INEGI 2010). Therefore, Portugal is a small country (approximately 92,000 km^2) with great wind power installed, which means that a large part of Portuguese mountainous ridges is occupied by on-shore wind turbines, greatly reducing natural areas. This impact can be very important for habitat specialists, such as as some raptors (Smith and Dwyer 2016), once the installation of wind turbines platforms, power lines, substations and access roads reduce available habitat, resulting in the loss of feeding, breeding, post-breeding and 'stopover' areas.

Such effects may be more drastic in small species, particularly in key places such as in feeding or breeding areas (Lindeboom et al. 2011). Normally, small species have small home ranges, which is a limitation when they need to find new places because it forces individuals to outlay a great energy expense. On the other hand, species that are more adapted to areas with human intervention usually are less affected than species adapted to natural or semi-natural areas (Pearce-Higgins et al. 2009). Human action, in terms of time (duration of interventions, e.g., establishing a wind farm) and also magnitude logically determine the extension of the impact.

A barrier effect is another behavioural impact identified, which is a consequence of the presence of a foreign element in a landscape that limits animals' free movement. In the case of wind farms, several observations suggest that some bird species prefer flying in areas away from turbines rather than inside turbine areas (Desholm and Kahlert 2005). Therefore, it can be hypothesised that all of these impacts, in the context of a wind farm, may induce a reduction of the use of the area by flying vertebrates, which might bring about a possible decrease in the risk of collision. On the other hand, in a small scale approach, habitat fragmentation itself means more 'areas to avoid' between vegetated patches, possibly inducing more movement among bats and birds between feeding sites, increasing collision risk. However, this barrier effect may even have other consequences. On one hand, an additional distance must be covered which may have an impact on the ability of individuals to conserve energy. On the other hand, wind farms can function as a barrier to local roosting and feeding flights, or to longer migratory flights (Fox et al. 2006).

2.4 Conclusions

Increasing numbers of wind farms seem to be inevitable given the international legal responsibility to reduce CO_2 emissions, but there remains much concern over the impacts on bird populations. With increasing numbers of wind farms comes concern not only over isolated environmental effects but also the cumulative environmental impacts. Despite awareness of the issue, there seems to be a lack of understanding and research in the area of cumulative impact assessment (Masden et al. 2009).

Reflecting the potential positive impacts on the earth's climate, the technological developments and associated reduced costs, wind power—along with solar power—has represented a huge economic investment worldwide. In favourable circumstances, i.e. with good resources and a secure regulatory framework, on-shore wind and solar PV systems also are cost-competitive with new fossil capacity, even without accounting for externalities (REN21 2016).

However, in parallel with the potential to mitigate climate change, wind energy is also known to have some negative effects on biodiversity, landscape and potentially in human health (Dai et al. 2015; Ryberg et al. 2013; Rydell et al. 2012), creating an antagonistic situation. However, it must be emphasised that the scientific community, environmental technicians, environmental entities, authorities and promoters have responded very satisfactorily. Therefore, efforts to find answers, solutions and mitigation measures have been made across the civilized countries. Because the solution for a better world, as in everything, is to find a balance that is reached with the openness and common sense of all involved.

References

Álvares, F., Rio-Maior, H., Roque, S., Nakamura, M., Petrucci-Fonseca, F. (2017). Ecological response of breeding wolves to wind farms: Insights from two case studies in Portugal. In M. R. Perrow (Ed.), *Wildlife and wind farms: Conflicts and solutions* (Vol. 1: Onshore. -Vol. 1: Onshore: 432, pp. 225–227). Pelagic Publishing.

APA. (2009). Guia para a Avaliação de Impactes Ambientais de Parques Eólicos.

APA. (2011). Guia Metodológico para Avaliação de Impacte Ambiental de Infra-estruturas da Rede Nacional de Transporte de Electricidade.

Arvizu, D., Bruckner, T., Edenhofer, O., Estefen, S., Faaij, A., & Fischedick, M. (2011). IPCC special report on renewable energy sources and climate change mitigation. by *O. Edenhofer et al. Cambridge University Press. Chap. Technical Summary*, 34–212.

Atienza, J.C., Martín Fierro I., Infante, O., Valls, J., & Dominguez, J. (2011). Directrices para la evaluación del impacto de los parques eólicos en aves y urciélagos (versión 3.0). SEO/BirdLife, Madrid.

Baerwald, E. F., & Barclay, R. M. (2011). Patterns of activity and fatality of migratory bats at a wind energy facility in Alberta, Canada. *The Journal of Wildlife Management, 75*(5), 1103–1114.

Baerwald, E. F., D'Amours, G. H., Klug, B. J., & Barclay, R. M. (2008). Barotrauma is a significant cause of bat fatalities at wind turbines. *Current Biology, 18*(16), R695–R696.

Bernardino, J., Zina, H., Passos, I., Costa, H., Fonseca, C., Pereira, M. J., et al. (2012). Bird and Bat mortality at Portuguese wind farms. In *IAIA12 Conference Proceedings' Energy Future The Role of Impact Assessment. 32nd Annual Meeting of the International Association for Impact Assessment 27 May–1 June 2012*, Centro de Congresso da Alfândega, Porto–Portugal.

Bourillon, C. (1999). Wind energy—clean power for generations. *Renewable Energy 16*1–4 , Jan, 1999: 948–953.

Canter (1996). *Environmental impact assessment* (2nd edn). McGraw-Hill, New York.

Chang, T., Nielsen, E., Auberle, W., & Solop, F. I. (2013). A quantitative method to analyze the quality of EIA information in wind energy development and avian/bat assessments. *Environmental Impact Assessment Review*, *38*,142–150.

COE. (2000). European landscape convention. *European landscape convention and reference documents* (pp. 5–20).

Cordeiro, A., Bernardino, J., Mascarenhas, M., & Costa, H. (2012). Long term survey of wind farms impacts on Common Kestrel's populations and definition of an appropriate mitigation plan. In NWCC Wind Wildlife Research Meeting IX. 27 a 30 de novembro de. (2012). *Denver.* EUA: Colorado.

Cordeiro, A., Mesquita, S., Marques, A. T., Silva, M. J., Rodrigues, N., Zina, H., et al. (2010). *Avaliação de impactes na componente biológica em parques eólicos: impactes previstos vs impactes reais* [Evaluation of impacts on the biological component in wind farms: Expected impacts vs real impacts]. CNAI 2010. 20–22 October, Vila Real, Portugal.

Correia, R., Faneca, C., Albuquerque, D., Vieira, J., Bastos, C., Pereira, M. J., et al. (2014). Characterization of the Ultrasonic Acoustic Field of a Wind Turbine. In *Conferencias y Comunicaciones del 45° Congreso Español de Acústica, 8° Congreso Ibérico de Acústica y Simposio Europeo de Ciudades Inteligentes y Acústica Ambiental. Murcia.* España. 29–31 October.

Cryan, P. M., & Barclay, R. M. (2009). Causes of bat fatalities at wind turbines: Hypotheses and predictions. *Journal of Mammalogy, 90*(6), 1330–1340.

Cryan, P. M., Gorresen, P. M., Hein, C. D., Schirmacher, M. R., Diehl, R. H., Huso, M. M., et al. (2014). Behavior of bats at wind turbines. *Proceedings of the National Academy of Sciences, 111*(42), 15126–15131.

Dai, K., Bergot, A., Liang, C., Xiang, W. N., & Huang, Z. (2015). Environmental issues associated with wind energy–A review. *Renewable Energy, 75*, 911–921.

Desholm, M., & Kahlert, J. (2005). Avian collision risk at an offshore wind farm. *Biology letters, 1*(3), 296–298.

De Lucas, M. M., Janss, G. F. E., Whitfield, D. P., & Ferrer, M. (2008). Collision fatality of raptors in wind farms does not the condor: Ornithological applications 118, 411–423. Q2016 Cooper Ornithological Society, Smith, J. A., & Dwyer, J. F. Energy infrastructure and birds 419 depend on raptor abundance. *Journal of Applied Ecology, 45*, 1695–1703.

DGEG. (2016a). Energia em Portugal—2014 [Energy in Portugal—2014]. Direção-Geral de Energia e Geologia [Directorate-General for Energy and Geology of the Ministry of Economy]. Available from http://www.apren.pt/fotos/newsletter/conteudos/energia_em_portugal_2014_dgeg_1459441498.pdf (in Portuguese).

DGEG. (2016b). Renováveis—Estatísticas Rápidas—n 145—novembro de 2016 [Renewables—Fast Statistics—nº 145—November 2016]. Direção-Geral de Energia e Geologia [Directorate-General for Energy and Geology of the Ministry of Economy]. Available from http://www.apren.pt/fotos/newsletter/conteudos/estatisticas_rapidas-renovaveis_novembro_2016_dgeg_1486118674.pdf (in Portuguese).

Drewitt, A. L., & Langston, R. H. (2006). Assessing the impacts of wind farms on birds. *Ibis, 148* (s1), 29–42.

Eggermann, J., da Costa, G. F., Guerra, A. M., Kirchner, W. H., & Petrucci-Fonseca, F. (2011). Presence of Iberian wolf (Canis lupus signatus) in relation to land cover, livestock and human influence in Portugal. *Mammalian Biology-Zeitschrift Für Säugetierkunde, 76*(2), 217–221.

EUROBATS. (2016). Report of the IWG on Wind Turbines and Bat Populations. 21st Meeting of the Advisory *Committee. Zandvoort, Netherlands, 18—20 April 2016.*

Fox, A. D., Desholm, M., Kahlert, J., Christensen, T. K., & Krag Petersen, I. B. (2006). Information needs to support environmental impact assessment of the effects of European marine offshore wind farms on birds. *Ibis, 148*(s1), 129–144.

Helldin, J. O., Jung, J., Neumann, W., Olsson, M., Skarin, A., & Widemo, F. (2012). The impacts of wind power on terrestrial mammals. *Swedish Environmental Protection Agency (Report 6510). Stockholm, Sweden.*

Hull, C. L., Stark, E. M., Peruzzo, S., & Sims, C. C. (2013). Avian collisions at two wind farms in Tasmania, Australia: Taxonomic and ecological characteristics of colliders versus noncolliders. *New Zealand Journal of Zoology, 40,* 47–62.

INEGI. (2010). *Parques Eólicos em Portugal*. Dezembro de 2010. p. 28.

Jay S. (2007). Customers as decision-makers: Strategic environmental assessment in the private sector. *Impact Assessment and Project Appraisal, 25*(2), 75–84.

Knopper, L. D., & Ollson, C. A. (2011). Health effects and wind turbines: A review of the literature. *Environmental Health, 10*(78).

Kowallik, C., & Borbach-Jaene, J. (2001). Impact of wind turbines on field utilization geese in coastal areas in NW Germany. *Vogelkundliche Berichte aus Niedersachsen, 33,* 97–102.

Morgan R. K. (2012). Environmental impact assessment: The state of the art. *Impact Assessment and Project Appraisal, 30*(1), 5–14. doi:10.1080/14615517.2012.661557

Larsen, J. K., & Madsen, J. (2000). Effects of wind turbines and other physical elements on field utilization by pink-footed geese (Anser brachyrhynchus): A landscape perspective. *Landscape Ecology, 15,* 755–764.

Leddy, K. L., Higgins, K. F., & Naugle, D. E. (1999). Effects of wind turbines on upland nesting birds in conservation reserve program grasslands. *Wilson Bulletin, 111,* 100–104.

Ledec, G. C., Rapp, K. W., & Aiello, R. G. (2011). *Greening the wind: environmental and social considerations for wind power development* (pp. 151). World Bank.

Lindeboom, H. J., Kouwenhoven, H. J., Bergman, M. J. N., Bouma, S., Brasseur, S. M. J. M., Daan, R., et al. (2011). Short-term ecological effects of an offshore wind farm in the Dutch coastal zone; a compilation. *Environmental Research Letters, 6*(3), 035101.

Loss, S. R., Will, T., & Marra, P. P. (2013). Estimates of bird collision mortality at wind facilities in the contiguous United States. *Biological Conservation, 168,* 201–209.

Marques, A. T., Batalha, H., Rodrigues, S., Costa, H., Pereira, M. J. R., Fonseca, C., et al. (2014). Understanding bird collisions at wind farms: An updated review on the causes and possible mitigation strategies. *Biological Conservation, 179,* 40–52.

Marques, A. T., Paula, J., Pereira, M. J. R., Ramalho, R., & Rodrigues, S. (2015). Assessing the problem. In M. Mascarenhas, J. Bernardino, A. Paula, H. Costa, C. Bastos, A. Cordeiro, A. T. Marques, J. Marques, S. Mesquita, J. Paula, M. J. Pereira, P. Pereira, F. Peste, R. Ramalho, S. Rodrigues, J. Santos, J. Vieira, & C. Fonseca (Eds.), *Biodiversity & wind energy: A bird's and bat's perspective* (pp. 30–51). Aveiro, Portugal: Bio3 and University of Aveiro.

Martin, G. R. (2011). Understanding bird collisions with manmade objects: a sensory ecology approach. *Ibis, 153,* 239–254.

Martin, G. R., Portugal, S. J., & Murn, C. P. (2012). Visual fields, foraging and collision vulnerability in Gyps vultures. *Ibis, 154,* 626–631.

Masden, E. A., Haydon, D. T., Fox, A. D., Furness, R. W., Bullman, R., & Desholm, M. (2009). Barriers to movement: Impacts of wind farms on migrating birds. *ICES Journal of Marine Science: Journal du Conseil, 66*(4), 746–753.

Paula, A., Marques, J., Pereira, P.S. & Santos, J. (2015). Chapter 4 - Mitigation: A Hierarchy of Solutions In Mascarenhas, M., Bernardino, J., Paula, A., Costa, H., Bastos, C., Cordeiro, A., Marques, A.T., Marques, J., Mesquita, S., Paula, J., Pereira, P.S., Peste, F., Ramalho, R., Rodrigues, S., Santos, J., Vieira, J. & Fonseca, C. (2015). *Biodiversity & wind energy: A bird's and bat's perspective, Bio3 and Univerity of Aveiro.* Aveiro, Portugal.

Pearce-Higgins, J. W., Stephen, L., Langston, R. H., Bainbridge, I. P., & Bullman, R. (2009). The distribution of breeding birds around upland wind farms. *Journal of Applied Ecology, 46*(6), 1323–1331.

Pimenta, V., Barroso, I., Álvares, F., Correia, J., Ferrão da Costa, G., Moreira, L., et al. (2005). *Situação populacional do lobo em Portugal: resultados do censo nacional 2002/2003* (p. 158). Instituto de Conservação da Natureza/Grupo Lobo, Lisboa: Relatório Técnico.

Piorkowski, M. D., & O'Connell, T. J. (2010). Spatial pattern of summer bat mortality from collisions with wind turbines in mixed-grass prairie. *The American Midland Naturalist, 164*(2), 260–269.

Piper J.M. (2001). "Barriers to implementation of cumulative effects assessment." *Journal of Environmental Assessment Policy and Management, 3*(4), 465–481.

REN21. (2016)—Renewables 2016 Global Status Report (Paris: REN21 Secretariat). ISBN 978-3-9818107-0-7.

Rollins, K. E., Meyerholz, D. K., Johnson, G. D., Capparella, A. P., & Loew, S. S. (2012). A forensic investigation into the etiology of bat mortality at a wind farm: Barotrauma or traumatic injury? *Veterinary Pathology Online, 49*(2), 362–371.

Rodrigues, L., Bach, L., Dubourg-Savage, M.-J., Karapandža, B., Kovač, D., Kervyn, T., Dekker, J., Kepel, A., Bach, P., Collins, J., Harbusch, C., Park, K., Micevski, B., & Minderman, J. (2015). *Guidelines for consideration of bats in wind farm projects—Revision 2014. EUROBATS Publication Series N. 6 (English version). UNEP/EUROBATS Secretariat* (pp. 133). Bonn, Germany.

Rydell, J., Engström, H., Hedenström, A., Kyed Larsen, J., Pettersson, J., & Green, M. (2012). *The effect of wind power on birds and bats – A synthesis. Swedish Environmental Protection Agency* (pp. 150).

Ryberg, J. B., Bluhm, G., Bolin, K., Boden, B., Ek. C., Hammarlund, K., Henningsson, M., Hannukka, I., Johansson, C., Mels,S., Mels, T., Nilsson, M., Skarback, E., Soderholm, P., Waldo, A., Widerstrom, I., & Akerma, N. (2013). The effects of wind power on human interests. *Swedish Environmental Protection Agency*, Report 6545, January 2013.

Schuster, E., Bulling, L., & Köppel, J. (2015). Consolidating the state of knowledge: A synoptical review of wind energy's wildlife effects. *Environmental Management, 56*(2), 300–331.

Smith, J. A., & Dwyer, J. F. (2016). Avian interactions with renewable energy infrastructure: An update. *The Condor, 118*(2), 411–423.

Sovacool, B. K. (2009). Contextualizing avian mortality: A preliminary appraisal of bird and bat fatalities from wind, fossil-fuel, and nuclear electricity. *Energy Policy, 37*(6), 2241–2248.

Thelander, C. G., Smallwood, K. S., & Rugge, L. (2003). *Bird Risk Behaviors and Fatalities at the Altamont Pass Wind Resource Area: Period of Performance, March 1998–December 2000* (No. NREL/SR-500-33829). National Renewable Energy Lab., Golden, CO.(US).

Chapter 3
Environmental Impact Assessment Methods: An Overview of the Process for Wind Farms' Different Phases—From Pre-construction to Operation

Joana Santos, Joana Marques, Tiago Neves, Ana Teresa Marques, Ricardo Ramalho and Miguel Mascarenhas

3.1 Introduction

3.1.1 Wind Farm Industry: General Overview

The need to tackle climate change, environmental pollution and to find sustainable methods to meet the increased demand for energy and power generation is critical and is being set out as primary goals by the European Union (Directive

J. Santos · J. Marques · T. Neves · M. Mascarenhas (✉)
Bioinsight, Odivelas, Portugal
e-mail: miguel.m@bioinsight.pt

J. Santos
e-mail: joana.s@bioinsight.pt

J. Marques
e-mail: joana.m@bioinsight.pt

T. Neves
e-mail: tiago.n@bioinsight.pt

A.T. Marques
Centre for Ecology, Evolution and Environmental Changes, University of Lisbon, Lisbon, Portugal
e-mail: ateresamarques@gmail.com

A.T. Marques
Centre for Applied Ecology "Prof. Baeta Neves", University of Lisbon, Lisbon, Portugal

A.T. Marques
REN Biodiversity Chair, CIBIO/InBIO Associate Laboratory, University of Porto, Vairão, Portugal

R. Ramalho
Bioinsight, Cape Town, South Africa
e-mail: ricardo.r@bioinsight.co.za

M. Mascarenhas et al. (eds.), *Biodiversity and Wind Farms in Portugal*,
https://doi.org/10.1007/978-3-319-60351-3_3

2001/77/EC, 27-09-2001). The European Directive 28/2009/EC introduces the obligation of the European Union (EU) member countries to submit a promotional plan for the use of energy from renewable sources. In the case of Portugal, the targets for the share of energy from renewable sources in sectors of transport (RES-T), heating and cooling (RES-a and a) and electricity (RES-E) have been defined in the Portuguese National Action Plan for Renewable Energy (PNAER). A national target of 31.0% of renewable in gross final energy consumption was defined to be achieved by 2020 (DGEG 2016).

Wind power is considered one of the fastest growing renewable energy sources worldwide, in part due to recent cost-competitiveness with conventional energy sources, technological advances, and tax incentives (Bernstein et al. 2006). By December 2015, the total installed wind-power capacity worldwide was approximately 432.88 Gigawatts (GW) with a new record of annual new added capacity in 2015 of 63.5 GW (GWEC 2016; WWEA 2016). Europe's total installed wind-power capacity by the end of 2015 was 147.77 GW (GWEC 2016). By this time, the country with the biggest installed capacity was Germany (44.90 GW) followed by Spain and the UK, with 23.03 and 13.60 GW, respectively (Fig. 3.1). Portugal follows with 5.08 GW (GWEC 2016). In 2015, 30.7% of the Electricity produced in Portugal came from a renewable source and specifically Wind energy contribution represented 22.5% of the country energy demand (APREN 2016), making Portugal the country with the second highest wind energy percentages on the electricity consumption (Denmark being the first). These results had led Portugal to be considered has an international example regarding the renewable energy sector.

Wind energy is, however, not free from environmental impacts (Drewitt and Langston 2006; Arnett et al. 2007; NRC 2007; Strickland et al. 2011; Arnett et al. 2016). The installation of wind energy facilities around the world has revealed issues regarding wildlife conservation (Eichhorn and Drechsler 2010), especially related to bird (Barrios and Rodríguez 2004; Drewitt and Langston 2008) and bat communities (Barclay et al. 2007; Arnett and Baerwald 2013; Arnett et al. 2016). Beyond birds and bats, habitat loss affects all existing biodiversity (Kikuchi 2008). Since 1992, when the first episodes of avian fatalities related to wind turbines were published (e.g., Orloff and Flannery 1992), social concern has arisen, and many articles and reports have been issued to date. To date, the potential for significant impacts remains a concern as many wildlife populations overlapping with wind energy development experience declines potentially caused by habitat loss, disease, non-native invasive species and increased mortality (AWWI 2015).

For Portugal, it is important to recognise the excellent results achieved by the wind energy sector in the country on the pursuit of the major goal of use renewable energy sources. Also important is the acknowledgement that wind energy is not free from impacts, which has led this sector's various stakeholders—developers, environmental authorities, specialized consultants—to the awareness of the challenge of guaranteeing the protection of the main biodiversity communities affected by wind farms while allowing the growth and development of the industry. Following this structure entails a strict impact assessment process and for that it is essential to comply with legal obligations and to ensure the application of adequate methods.

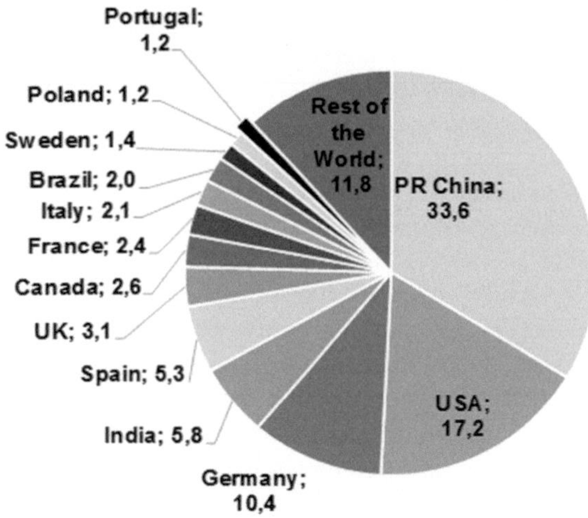

Position	Country	MW	% Share
1	People's Republic of China	145 362	33,6
2	USA	74 471	17,2
3	Germany	44 947	10,4
4	India	25 088	5,8
5	Spain	23 025	5,3
6	UK	13 603	3,1
7	Canada	11 205	2,6
8	France	10 358	2,4
9	Italy	8 958	2,1
10	Brazil	8 715	2,0
11	Sweden	6 025	1,4
12	Poland	5 100	1,2
13	Portugal	5 079	1,2
	Rest of the World	50 947	11,8
	Total Top 13	**381 936**	**87,1**
	World Total	**432 883**	**100,0**

Fig. 3.1 Graphical stats for the 13 world countries wind energy installed capacity in 2015: total production (MW—megawatt) and global share in a worldwide scenario (percentage). Adapted from GWEC (2016)

This chapter aims to address the best practices that are followed in Portugal to comply with the mentioned legal obligations and adequate methods complemented with the analysis of good examples in other countries.

3.1.2 Legal Requirements and Guidelines

In general, assessing the environmental impacts of a project is a legal obligation for developers in Europe, based initially on the European Directives 2001/42/EC and

85/337/EEC (European Commission 2011), which later were incorporated in most European countries' national legislation. When planning a new project, developers first have to validate its relevance in terms of social, economic and/or the environmental compliance requirements of the proposed infrastructure. In this process, several environmental descriptors have to be analysed, including biodiversity (i.e., ecology), water resources, soil or landscape. During this evaluation, developers have to demonstrate that the project will not have significant impacts on biodiversity (i.e., habitats and species).

An adequate legislation together with the good will of all the enforcing stakeholders is fundamental in the definition and evolution of the methodologies used in the impact assessment. Good available tools to assess this issue include several national guidelines. The basis is either to orient developers on how and where to install new wind farms and how to mitigate impacts, or consultants on proposing and implementing monitoring programmes with standard experimental designs.

The main goal of the Environmental Impact Assessment (EIA) process is to guarantee that all negative impacts are identified in an early stage of the project and that they are eliminated, mitigated for and, ultimately, compensated (if no alternative is possible), following the mitigation hierarchy (BBOP 2012). Regarding wind farms, the main impact receptors, in terms of biodiversity, are bird and bat communities, but also other groups such as mammals and vegetation (Flora and Habitats) are known to suffer negative impacts (e.g. Kuvlesky et al. 2007; Santos et al. 2010) comprised in the EIA process.

In the Portuguese case, wind farms, especially those located within or in the vicinity of classified areas (e.g., national and natural parks, natural reserves,), are subjected to a strict EIA process. In this respect, the Portuguese authorities have been requiring (with regards to the ecology descriptor for the EIA) a dedicated bat and bird study for the proposed development area, sometimes covering a full annual period before the wind farm development can be approved. One of the first cases where this strategy was applied was the Chão Falcão II Wind Farm, in 2004–2005, where both fauna and flora communities were studied over a year (Bio3 2005). Since then, approaches have been refined focusing on the groups that are known to suffer more significant impacts (birds and bats), being some examples Bornes Wind Farm in northern Portugal (Ecosativa 2007; Plecotus 2008), Prados Wind Farm in central Portugal (Bio3 2011) and Mértola Wind Farm in southern Portugal (Bio3 2012), with similar requirements for pre-construction phase. The Portuguese Environmental Agency (http://www.apambiente.pt/) provides instruments to develop the EIA processes, including the specification of legal framework, guidelines and recommendations of entities to involve during the process. In addition, during the operation phase, it is mandatory for several wind farms to implement biodiversity operational phase monitoring programmes covering several years on the cycle of the wind farm and periodic reviews of the plans when necessary (e.g. Bioinsight 2016a, b, c). For the last 20 years, stakeholders have acquired a significant knowledge regarding assessment of impacts of wind farms on biodiversity, which has been supported by a good amount of studies and operational phase monitoring programmes. Of the 259 wind farms in operation in Portugal by

Table 3.1 Specifications of the Portuguese legal framework for EIA concerning **wind power projects for electric production**.

General cases		Sensitive/Protected areas/Natura 2000	
Mandatory	Project typology	Mandatory	Project typology
EIA	WF ≥ 20 turbines or located near (<2 km) to other similar projects	EIA	WF ≥ 10 turbines or located near (<2 km) to other similar projects
EIA	Repowering/Upgrading (+turbines) of WF already in operation that have not been previously subjected to EIA process, if the final layout of the WF project sums 20 or more turbines or if the distance to other similar projects ends up being <2 km	"EIncA"	All WF within the areas that are not covered by the above specification

Situations where EIA and 'EIncA' are mandatory. Adapted from the Law Decree No. 131B/2013 of October 31st, amended by Law Decree No. 179/2015 of August 27th

December 2015, the majority of them were under monitoring programmes covering the different project development phases.

The following paragraphs summarize the best methods adopted on impact assessment in Portugal, in particular the best practices for biodiversity survey and monitoring. Although the main focus is Portugal, it is important to make an overview on the differences and similarities of guidelines for other countries.

In Portugal, the basis of the environmental policy is defined in the Law 19/2014 of April 14th. More specifications are declared by Decree Law No. 131B/2013 of October 31st, amended by Decree Law No. 179/2015—Official Gazette no. 167/2015, 1st Series, of August 27th, which establishes the legal procedures of EIA for *public and private projects likely to have significant effects on the environment*, transposing into national law the European Directive 2014/52/EU of the European Parliament and of the Council of December 13th 2011 on the assessment of effects of certain public and private projects on environment.

Since wind farms are included in the above-mentioned legal documents, the associated EIA is ruled by the Portuguese legal framework. The requirement for a wind farm to be subjected to a full EIA process depends of the project's proposed location and power production (Table 3.1). The Portuguese legal framework also defines the targets that have to be fulfilled when implementing an EIA:

(a) Evaluate, through an integrated approach, the potential significant direct and indirect impacts that will result from the implementation of the Project and the proposed alternatives (e.g., layout configuration) in order to support the decision on the environmental viability of that same Project;
(b) Define the measures to avoid, minimize or compensate those impacts, supporting environmentally sustainable decisions;
(c) Establish the post hoc verification process to evaluate the effectiveness of any enforced measures, always including the monitoring of the real effects of the assessed Project;

(d) Ensure public participation and consultation of all stakeholders to formulate decisions that concern them, emphasizing dialogue and consensus during the performance of the administrative functions.

The legal framework that regulates licencing of electrical production based on renewable energy in Portugal also extends the need of environmental studies to wind energy development projects that are not covered by the specifications that require an exhaustive EIA. These situations are ruled by the Law Decree No. 225/2007—Official Gazette no. 105, 1st Series, of May 31st, where all projects to be developed in Protected, Sensitive or Natura 2000 areas are necessarily subjected to an environmental assessment (called "Environmental Incidences Assessment—EincA"), regardless of the projected number of turbines/power production (Table 3.1). The basis of these studies are similar and aim for the same general targets of the EIA, although depending of the project they may be less exhaustive and focus more on specific environmental characteristics of the sensitive area. Nonetheless, since the impact assessment procedures are almost equivalent to the EIA, including studies during all the phases of the project development, in this chapter henceforth, EIA and "EIncA" are not differentiated.

To complement these legal requirements, the Portuguese Environmental Agency produced, in 2010, a list of guidelines supporting the EIA processes specifically for wind farms, providing a guidance for the best practices to the complete timeline of a given project (APA 2010)—including the scoping and environmental impact studies, pre-construction, construction and operation phases of a project. Among these practical guidelines, one can highlight (on the matter of the assessment methods) the definition of monitoring plans, mitigation measures (minimisation and compensation), and focus on the main impact receptors: birds and bats. Portuguese guidelines also include specifications for large carnivores (i.e., the endangered Iberian wolf, *Canis lupus*) as well as flora and habitats.

The general guidance and recommended procedures at the national level are well within other well established international guidelines. Some comparative examples can be not only the remaining Iberian Peninsula (Spain), but also the northern Europe (across the United Kingdom and Scotland in particular), North American (USA) and African (South Africa) procedures. A compilation of the major procedures defined in guidelines in the referred countries is presented in Appendix (Table 3.5), where is clear that general documents aiming to define wind farm impact assessment focus mainly only the two groups known to be most affected—birds and bats. One thing of note is that none of the analysed guidelines considers the impact of wind farms on carnivore populations—for example, wolves. This includes Spain, where this animal is present and is considered an endangered species (Blanco and González 1992).

Regarding birds and bats, almost all the compared guidelines recommend a before-after-control-impact (BACI) approach to assess the impacts of construction and operation of a wind farm. Methods to assess this impact in birds also have similarities in the analysed documents, recommending assessing common population parameters according to species phenology, behaviour, ecological features in the study area. Although some differences may arise when analysing the depth of specifications that are made (Table 3.5, see Appendix). The same situation can be found for bats. For both groups, fatality assessment surveys arise as one of the major issues to address during operation phase of a wind energy project, although different specifications are found. For example, in the USA, where wind farms have usually many wind turbines, it is recommended that only 30% of them must be searched, while, for example, in Portugal, searches must be conducted on all wind farms or in at least 70% of the wind turbines.

From a comparative analysis of the national and international guidelines, one can see that the duration of the monitoring programmes varies between countries. In particular, for the construction phase of wind farms, when negative impacts may occur, several guidelines do not address this period. Portuguese guidelines recommend monitoring during construction, generally full coverage monitoring. For the other phases of project development (pre-construction and operation), the duration of monitoring programmes in Portugal is usually recommended to be one year in the pre-construction phase and three years in the operational phase, after construction has ended. These recommendations are the most common but not limited since, in some cases, monitoring may be required for the entire life cycle of the project. Monitoring plans should be reviewed annually and adjusted every time they are needed. These durations are not applied for flora and habitats but are applied for bats, birds and large carnivores. Some countries do not have specific durations for monitoring programmes and others differ greatly from the Portuguese guidelines.

No standard guidelines or experimental designs are defined to be followed worldwide, not only due to the countries legislation regarding wind energy assessment needs but also its position in investigating and understanding the national reality regarding biodiversity and wind farm interactions. Independently of specifications given in particular guidelines, the impact assessment scheme follows the same general approaches in Portugal as in other countries, as explained next.

3.2 Impact Assessment Methods

With the awareness that wind energy development may have negative consequences over wildlife, the need for identifying the real impacts has arisen as a major challenge to specialists. In recent decades, efforts have been made through intensive research to understand the full dimension of the problem, which can only be achieved through the implementation of standard sampling methods, and the development of new methodological approaches to this field.

By analysing the different guidelines produced both national and internationally, one can easily find that there are several common factors among them regarding the methods of impact assessment:

- A detailed environmental study/report when planning a new project is a crucial step of the EIA process, since the evaluation focuses on predicting the potential resulting impacts that the project may be responsible for (potential impacts);

- Accurate monitoring (and mitigation if necessary) plans during construction and operation phases must allow for:
  - Measuring the extension of the effects predicted in the EIA report or Environmental Study (ES) phase and/or identifying any effect that had not been predicted (real impacts):
    - Standard methodology should be adopted to allow comparison between sites, by means of scientific methods;
    - Survey and monitoring studies should have the correct duration to allow for proper data collection and, consequently, data analysis for impact validation;
    - Data analysis should be robust enough to allow statistical significance analysis.

The assessment methods to determine impacts should consider both pre-construction and post-construction phases of the wind farm. Only by focusing on the overall process and life cycle of the wind farm can one determine with accuracy if the real impacts are according to the predicted impacts (EIA report development phase) or, if they are not, why and what are the consequences on biodiversity. A continuous evaluation process must also lead to the identification of the need to adapt the monitoring plans, target species and mitigation techniques.

3.2.1 *Planning and Project Phase: Evaluation of the Potential Impacts*

3.2.1.1 Scoping

Scoping consists of producing a scoping report in an early stage of the project planning and design development. In Portugal, this phase aims to assist the developer to identify the environmental factors present in a proposed area that will have to be further investigated during the EIA process, similar to international proceedings (Sea Green Wind Energy 2011; ABO Wind UK 2014; SNH 2014—Table 3.5, see Appendix).

The main work consists of collecting data leading to an adequate selection of the preliminary area to implement the project, if several options are being studied. This procedure is the first and best way to mitigate, or greatly reduce, the impacts of wind developments on biodiversity (Marques et al. 2015). Information gathering is mainly based on legislative requirements, bibliographic information, international conventions and expert knowledge. Being conducted in an early phase, this preliminary evaluation may allow developers to avoid unnecessary pre-development investment in lands not suitable for development, or to promote meetings to discuss any concerns, including environmental issues.

Mostly, desktop survey is conducted to identify the most sensitive areas and the main potential conflicts for wildlife communities that should promptly be incorporated in the planning of the project and EIA. For wind farms, special attention is given to birds and bats, as previously stated. These target groups are perceived as more susceptible to impacts due to eco-morphological and behavioural characteristics that may increase collision risk (Barrios and Rodríguez 2004; Horn et al. 2008; Santos et al. 2013; Marques et al. 2014; Shuster et al. 2015). In Portugal, the scoping report also addresses large carnivores due to potential conflicts with the endangered Iberian wolf, *C. lupus* (if the project location is within the distribution range of the species), as well as flora and habitats (preliminary assessment of the potential presence of RN2000 areas from Habitats Directive and/or important plant species).

3.2.1.2 EIA Report or Environmental Study

The EIA process starts upon completion of the scoping phase. Detalied studies are conducted to ensure that all relevant themes, background information and timelines are identified at the earliest stage of a development proposal. This process also includes development of further guidance as to the level of effort required to adequately assess impacts.

In this phase, an Environmental Study (ES) is prepared. Generically, the ES consists of the evaluation of impacts and alternatives, i.e. to identify the potential environmental impacts of a proposed project, including the detailed study of different layout alternatives. Specifically, the main purposes are to report the configuration, layout of a given project or development (including alternatives, whenever applicable); to identify the main probable negative impacts and impact receptors, their significance, how to avoid them; and, finally, to propose accurate mitigation and offset/compensation measures for unavoidable impacts. Following the mitigation hierarchy (1. avoid; 2. minimise; 3. rehabilitate/restore; 4. offset/compensate) (BBOP 2012) at this stage is crucial and is acknowledged in Portugal as well as internationally (ICNB 2010a; Paula et al. 2015a; Peste et al. 2015; Barré et al. 2015) to achieve "no net loss" or even "net gain" in biodiversity (BBOP 2013).

In content matters, the report should document and map all natural heritage features, areas and their ecological functions (including target group species) that are expected to be affected, on or adjacent to the proposed development. In addition, the ES should describe and assess the diversity and connectivity of natural features within the project area and, where applicable, at a broader scale to maintain, restore or improve the long-term ecological function and biodiversity of the natural heritage system.

During this step, intensive baseline data collection should be undertaken, both desktop and fieldwork, aiming for a detailed characterisation of the wildlife and/or relevant target communities of an area (Marques et al. 2015). In Portugal, although not legally mandatory, the minimum recommended survey period during project planning phase is one year (surveys covering the four seasons of one year) by the Portuguese Environmental Agency (APA 2010), at least for fauna communities.

Currently, most of the developers undertake this methodology. Although the recommended pre-construction survey period is inferior when compared with other examples in northern Europe, it is in accordance to South African and general methods for Iberian Peninsula (Spain) (Table 3.5, see Appendix). For vegetation and natural habitats, baseline information is usually required, although fieldwork and survey period is not specified in Portuguese guidelines. Nonetheless, this is a theme that usually is not referred to in international guidelines.

Provisions for a follow-up programme need to be made in the pre-decision EIA process and carried out post-decision. Cost-effective evaluation strategies that can be carried out throughout different phases of project development are greatly significant. Thus, the baseline data collection methods (specifically pre-construction monitoring) should be (if possible) consistent with the monitoring programme set for the construction and operational phases, while acting as reference data, which is in accordance with the guidelines. That is to say that the monitoring plans between different phases should be designed following a BACI analysis (Smith 2002).

The results achieved during baseline collection are included in the overall EIA report and significantly contribute to the identification of the expected impacts and the need to mitigate them. Final recommendations from the report may include additional study requirements, restoration and the need of the review of monitoring programmes according to any changes that are made throughout the follow-up process. The strict analysis of the ES by the national environmental authorities and considering both expert and public consultation inputs and final decision-making is later formalized by the outcome of an *Environmental Impact Statement*. Here, it is stated whether the project is legally authorized and, if so, if there are any conditions that the developers are obliged to follow. In Portugal, every measure stated in the Environmental Impact Statement represents legal obligations for developers, and they are organized by project phase: pre-construction, construction, operation and deactivation (if applicable).

Finally, it is important to underline that both in Portugal and internationally any measures and monitoring programmes defined at this stage are not assumed to be static. Rather, they should be open to improvement or reviews during the construction and operation phases of the wind farm, according to the knowledge and response of wildlife populations to the project by means of an adaptive management approach (Williams 2011; see Coelho et al. Chap. 8).

In summary, the conclusion of this Planning and Project phase should result in the following outputs:

- Adequate selection of the area to develop the wind farm (project final layout definition by developers)—best solution: less potential impacts;
- Definition of the best measures to mitigate, or greatly reduce, the impacts of wind developments on target wildlife communities (if necessary);

- Proposal of monitoring plans to implement in the next phases (including real impact assessment and mitigation effectiveness evaluation).

About Cumulative Impact Analysis During EIA

The importance of evaluating cumulative impacts

Several countries such as Portugal have developed a long-term strategic plan for renewable energy. This implies the construction of several wind farms during several years, creating large areas occupied by wind farms. One wind farm may not cause significant impacts but two or more projects located in the same area or region may have a cumulative effect and cause a significant impact on the local population of bird and bats.

The difficulty of conducting that evaluation

Evaluating cumulative impacts in an EIA process is difficult, either because of the lack of information, or shared information about the future of wind farms or because of the difficulty of executing or even agreeing on the methods to be used. In Portugal, for example, legal frameworks and general guidelines for impact assessments refer to the importance of proceeding to a cumulative analysis, but no methods or standard procedures have been proposed to date.

In the Scottish Natural Heritage (SNH) guidance of cumulative impact assessment (2012), scenarios which assume 100% disturbance within a pre-determined distance of turbines can be derived for key species using conservative threshold disturbance distances (Whitfield and Ruddock 2007, as cited in SNH 2012). Empirical evidence is lacking for most species but some indication of real displacement distances from wind farms can be taken from Pearce-Higgins et al. (2009). Even more specific is an approach by Bastos et al. (2016) who made a spatially explicit dynamic analysis combining the results from modelling techniques to study, by means of simulations, the cumulative impacts of wind farms on the skylark, *Alauda arvensis*, at a regional level in northern Portugal where several wind energy development projects are currently planned or in operation. Although these studies are site- and species-specific they may provide good baseline information for further investigation and data extrapolation.

3.2.2 Construction and Operation Phase (Evaluation of the Real Impacts)

Monitoring actions must be considered to be a crucial element of the EIA process (Marshall et al. 2005). When considering wind farms' construction and operation phases, monitoring programmes must provide information to evaluate whether the predicted impacts occur as defined in the previous phase and, if not, the identification of the real impacts including new ones, not initially predicted. Equally important is to check whether the proposed mitigation measures are proceeding as planned, to identify any failed mitigation measures and, if so, to address them in the shortest possible time.

To achieve these goals, it is crucial to define the best strategies that will assure the correct implementation of every measure described in the Environmental Impact Statement, including the fulfilling of monitoring plans. Monitoring programmes should be designed to produce accurate results and conclusions, according to: (i) the target groups/species; and (ii) the objectives of the assessment—also related to the determination of the level of effort to apply in the study (see Appendix—Table 3.5). It is important to underline that there are no standard procedures to the experimental designs: each construction/operation programme needs to be site-specific, since each has its own environmental characteristics, vegetation types, and animal species cast. So, they must be individually evaluated. Nonetheless, various methods are offered to assess the real effects of a wind farm according to the target groups (or species) and the type of predicted impact.

Natural populations are complex research subjects, with intricate relationships to many other biotic and abiotic key elements on the study area. Monitoring programmes using the BACI design will allow the real impacts of a wind farm to be distinguished from natural variations and stochastic events that occur in natural communities (Smith 2002; de Lucas et al. 2007).

3.2.2.1 Fauna

Negative impacts usually expected on the main fauna impact receptors are among the key factors assessed throughout monitoring programmes. As detailed by Pereira et al. (Chap. 2), the known negative impacts include displacement, disturbance, barrier effect and direct mortality due to turbine collision (the last one only during the operation phase and just for flying vertebrates) or barotrauma (in the specific case of bats). Since habitat loss is a negative impact that is considered to be negligible, and due to the difficulty to distinguish it from displacement and disturbance, it is usually not assessed during the monitoring programmes.

Here, the main goal is to understand if the predicted impacts during the ES corresponded to the real impacts of a project and to define the tools to mitigate them, whenever necessary.

To assess the occurrence of impacts such as disturbance and/or displacement over wildlife populations, it is necessary to study the occurrence of fluctuations in population numbers and/or in the way communities use a certain area, which will arise from changes in species compositions, demography, habitat use, behaviour, among others (see also Schuster et al. 2015 for an extended bibliographic review). During the operation phase, caution is needed when analysing such variations, since they might also result from direct fatality, especially when monitoring bat and bird populations. The abundance of a certain species might decrease within the area of a wind farm and this can be either due to displacement or due to a high fatality rate. Therefore, fatality needs to be considered not only specifically but also when analysing other negative impacts.

Monitoring programmes also have to consider the fact that natural populations may be affected by different driving forces, such as inter-species competition, major habitat alterations or even climate change. This is even more important when dealing with migrant species—both birds and bats—that may depend on different geographic areas to fulfil their life cycle (reproduction, foraging and winter season may occur in different places, even different continents).

This is the reason why programmes should follow BACI experimental designs, since the comparison with baseline years and the with fluctuations observed in control areas, results in the assessment of the existence of (Marques et al. 2015):

- Changes in the habitat use patterns by wildlife communities (including air space for bats and birds);
- Differences in species or communities' behaviour due to the wind farm presence;
- Changes in the species composition (species richness; diversity) in or near the wind farm;
- Differences in the abundance patterns (density; number of individuals) of the communities or certain species in or near the wind farm;
- Changes in the breeding performance (production of offspring/fledglings)—mostly relevant if there are endangered target species.

Various field methods and techniques are available to assess the above-mentioned evaluation. As shown in the Appendix, Table 3.5, the methods vary among countries according to specific guidelines. Nonetheless, one can summarize the main field techniques available for collecting data depending on the main target groups in wind farm impact assessment for fauna communities (Table 3.2). Usually, bats and birds are monitored separately, but there are some cases when it is possible to combine sampling efforts.

In Portugal, standard survey methods focus on indirect sampling (point count/transects) for the estimation of activity/abundance indexes (APA 2010; ICNB 2009, 2010b,c; Bioinsight 2016a, b). This is an approach that allows answering the main goals of a wind farm construction/operation monitoring programme with a good cost–benefit relation. For example, birds are usually surveyed by visual (direct observation) or acoustic census (singing birds) depending on the target communities (e.g., Bibby et al. 2000; Hardey et al. 2006); bats (echolocating species) have been typically surveyed through acoustic detection via ultrasound devices (Parsons and Szewczak 2009; Britzke et al. 2013); and for large carnivores, sampling techniques such as surveys of animal signs (e.g., dungs, footprints) and remote camera trapping are known to provide information on the species presence and distribution in an area (Gese 2001). On the theme of cameras in wildlife monitoring, it is also important to highlight the use of infrared cameras in bat surveys to assess species activity near wind turbines (Horn et al. 2008), although this is not a common approach in Portugal and is not found in national guidelines (see Appendix, Table 3.5).

However, if the aim is to analyse bird or bat flying routes, standard methods may not be robust enough to provide specific information. In this field, other technologies adapted to ecological features such as radar systems or GPS/radio tracking methods can be used (Table 3.2) to assess specific behavioural patterns, as well as distribution of other ecological patterns (Kunz et al. 2007). Radar tracking is more

Table 3.2 Field methods available to collect data for assessing disturbance and/or displacement impacts on fauna communities during construction and operation phases in WF

Field method	Target group/Species	Description
Visual/Acoustic census	Birds	Point or transect counts. A human observer recognises and registers birds by their call and/or by sight and morphological traits, quantifying birds' numbers. Notes on behaviour, flight height, direction, etc. can be made; flying routes can be drawn on a map to allow detailed spatial analysis in GIS environment
Acoustic detection	Bats (echolocating species)	Point or transect counts. Bat presence is determined when pulses are detected by an ultrasound device. When recorded, these pulses can be used to identify species, at high level of certainty (species or group of species). Manual equipment is used to sample near the ground. Automatic devices can be placed at different heights
Radar	Birds and bats	Radar equipment can collect detailed data on flying birds and bats, on parameters such as flight direction and speed. The analysis of the target's dimensions and behaviour can provide information on the species
Mist-netting	Birds and bats	Mesh nets suspended between two poles are used to catch individuals flying by. Individuals are identified by means of morphological characters and a distinct ID code is given
Tracking	Birds, bats and large carnivores (Iberian wolf)	Tracking devices such as ultra-light radio transmitters/GPS loggers/PTT (Platform Transmitter Terminals) can track individuals along the area, mapping their routes. This technique involves the capture/release of the animal to which a transmitter is attached
Camera trapping	Large carnivores (Iberian wolf); Bats (less common)	Point location of digital infrared cameras. Non-invasive method is a powerful tool to ecology monitoring studies for mammals. Provides information about species distribution, abundance and behaviour. Capture–recapture models can be used for individual differentiation and to estimate population numbers
Indirect counts/Presence of signs	Large carnivores (Iberian wolf)	Transect counts. Non-invasive method where the observers search and count signs of presence of a species, such as dungs, in a linear transect. Other signs of presence, e.g., footprints or predation signs can also be recorded. Data analyses allows determination of species occurrence, distribution and abundance. Fresh dung samples can be collected for genetic analysis to provide biological information at a population level

Adapted from Marques et al. (2015)

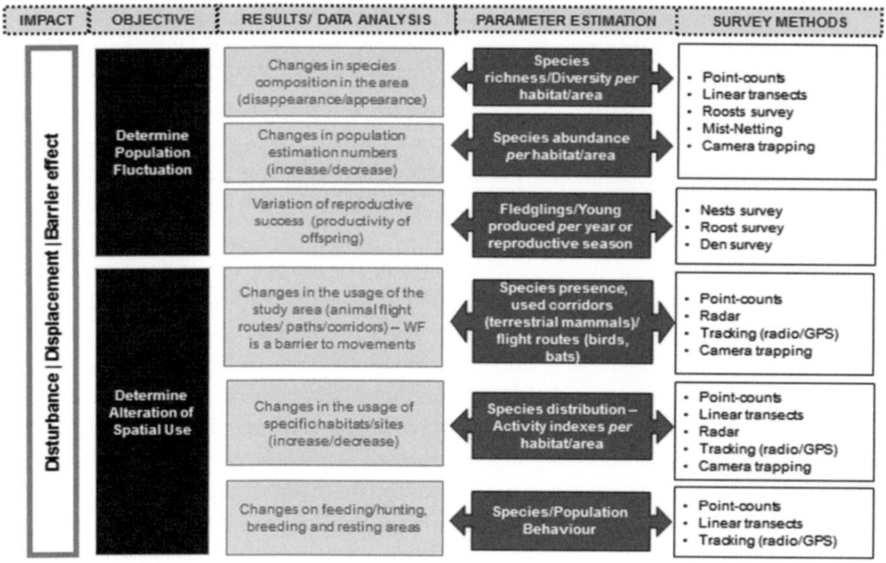

Fig. 3.2 Summary of commonly used methods to assess disturbance/displacement impacts during wind farm post-construction phase. Adapted from Marques et al. (2015)

common in bird studies (Fischer et al. 2012; Fijn et al. 2015) due to the ecological traits of this group (animal size and flight patterns among other factors). Depending on the technical specifications, it may be a good tool to provide data not only on general bird routes but also to identify species or groups of species detected through the application of machine learning algorithms (Rosa et al. 2016). The use of radar in ornithology studies in wind farms is also an adequate tool when dealing with mitigation measures, since it provides powerful information into turbine shutdown on demand when monitoring bird activity in wind farms aiming to reduce mortality due to collision (e.g. STRIX 2015). See also a more complete analysis of both bird and bat monitoring examples in using radar systems in Portugal in the 'case study' section of this chapter.

According to the objectives of each study and by using the adequate sampling methods depending on the target species, a set of monitoring parameters can be determined and its results can be analysed per se or combined (Fig. 3.2).

After the determination of the population/species monitoring parameters, the identification of a specific impact is still not a straightforward task and a careful analysis is required (Drewitt and Langston 2006). By comparing changes in this scenario using the BACI method, and applying statistical analysis, it may be determined whether the target species/group is being influenced by the wind farm construction or operation. Usually, the more accentuated the difference the greater the extent of the impact.

In the specific case of migrating species or soaring birds, the barrier effect becomes evident when routes are analysed and species change their natural path to avoid the presence of the wind farm that acts as a barrier (Masden et al. 2009). However, other negative impacts are not always possible to isolate and identify, since they can act together in the same population and exhibit the same result (e.g. displacement/disturbance effect). For example, birds or bats can avoid an area completely (total exclusion), can be present in the area but in decreased numbers (displacement) and the remaining population can still be the subject of disturbance effects with indirect consequences on the population, such as a decrease on the reproduction success.

Finally, the identification of these impacts cannot be dissociated from the fatality monitoring (see next paragraphs), as the described decreasing population can either result from fatality or displacement and disturbance effects. Given the example of migrating species, if the species does not display changes in activity and flight migration route patterns in a wind farm development area, this behaviour combined with other environmental factors may result in a high collision risk (Horn et al. 2008).

Direct fatality due to collision in wind farms is a negative impact that constitutes one of the core questions to assess in post-construction monitoring programmes, particularly concerning birds and bats since non-flying vertebrates are not subject to this negative impact. Understanding the mortality rates in which both groups are dying in consequence of the wind farm operation is usually a key issue of studies of environmental impact assessment (e.g., Barclay et al. 2007; Drewitt and Langston 2008; Kikuchi 2008; Marques et al. 2014). However, this is not a simple question in real impact assessment. In fact, there are many methodological details to consider in the assessment that do not have a standard procedure (Marques et al. 2015), namely:

- Definition of a carcass search protocol to find flying vertebrates killed at wind turbines, including survey techniques and frequency of the sampling design;
- Identification of potential bias to consider to the carcass searches and findings:
  - what affects the detectability of carcasses?
- How fast are carcasses removed by scavengers or decomposed in the wind farm?
- What other factors may affect or bias the sample?

When a fatality occurs, carcasses normally stay in the vicinity of the wind turbines, thus, currently, researchers perform periodic carcass searches to detect and quantify collision events, which is used to estimate fatalities. Although not mandatory by law, monitoring fatalities in Portugal is well defined in the Portuguese Environment Agency guidelines for monitoring programmes compared with other countries (Table 3.5, see Appendix). The frequency of the searches should be defined on a case-by-case scenario; however, it specifies that, for birds, searches must cover the 12 months of the year and, for bats at least 8 months in a year, according to the known activity patterns in Portugal for this last group (ICNB 2010c).

Working in wild environments has several methodological uncontrollable constraints. Examples include vegetation cover (that may compromise the effective search area), the climate conditions and scavenger populations of each region. These conditions need to be taken into account when estimating bird and bat fatalities, since:

- After collision, carcasses start to decompose or may be consumed as nourishment by scavenger species. If not properly included, this factor will underestimate the fatality rate;
- Another source of bias is the fact that observers performing carcass search trials do not find 100% of the carcasses featured on the searched area.

Thus, to improve fatality estimates, researchers usually include at least two correction factors: *carcass removal* due to scavenging or decomposition and *searchers efficiency*, both of which are subject to field trials. Table 3.3 summarises the goals to achieve on this matter in a monitoring programme and the typical experimental approaches being used in Portugal.

Field methods traditionally rely on human observers to perform both the searches and the trials (Bernardino et al. 2012) but, in recent years, investigations have been made into the efficiency of field methods, from a cost-efficiency perspective. In the specific case of carcass removal trials, recently, in Portugal, digital infrared cameras were used to evaluate carcass removal rates in wind farms (Paula et al. 2015b). In the case of carcass searches, surveys performed by humans are known to be a method with relatively low efficiency and accuracy and highly dependent on the carcass size (for birds) and vegetation structure (Smallwood 2007). As an alternative, in the last decade, the use of specially trained dogs to perform carcass searches in wind farms has been studied with results revealing significant improvements on carcass detection when compared with humans (Arnett 2006; Mathews et al. 2013). In Portugal, this approach is also an acknowledged method to minimise the influence of detectability in the mortality estimates, as showed by Paula et al. (2011a). Although this still not a widely used approach, this method is currently used in some monitoring programmes regarding carcass searches and fatality estimation in Portugal (e.g., Prados wind energy facility in Serra da Estrela, Portugal; Bioinsight 2016a); see also Paula et al. (Chap. 6).

Finally, is also important to account for another potential source of bias that is related to the place where the carcasses fall and the surveyed area during the carcasses searches. Studies found that a significant portion of the birds and bats carcasses fall outside of the usually sampled areas (Hull and Muir 2010; Huso and Dalthorp 2014). To avoid sampling a large area that may not be a cost-effective approach or that may not even be possible to implement in the field, one option is to obtain a bias correction factor through mathematical modelling (Hull and Muir 2010). However, currently, monitoring programmes in Portugal do not include this correction factor in the experimental designs for fatality estimation, focusing on the major correction factors for carcass removal and search efficiency rates.

Through the combination of the results from the carcass search with the values of the different correction factors, it is possible to estimate an approximation to the real fatality numbers resulting from wind farm operation. These estimates result

Table 3.3 Field methods to collect data to determine two correction factors aiming fatality estimation of birds and bats during wind farm operation

Parameters	Objective	Description
Carcass searches	Detect and quantify collision events in a wind farm	Carcass searches are surveyed by one or two field technicians. Every turbine is searched under the rotor blades to find collision victims. Once a carcass is found, the observer proceeds to the identification of the species and, if possible, parameters such as age, sex and approximate time and cause of death. Searches are performed periodically according to the target groups/species and the characteristics of the study area. The quantification for the observed mortality is then used to estimate the real fatality in the wind farm, by applying the correction factors through mathematical formulae
Carcass removal rate	Assess carcass removal values according to the characteristics of the wind farm area	Carcass removal trials are performed by one technician throughout periodic visits to previously known carcasses, in order to check how long their permanency is detectable and to identify the changes due to decomposition or their disappearance. The removal of carcasses is site-specific and is dependent on weather conditions, and local community of scavengers, among other factors
Searcher efficiency rate	Determine the observer accuracy in finding carcasses according to the characteristics of the study area	Carcass search trials are performed by or more field technicians (the presence of the technician that performs the carcass periodic searches in the wind farm is mandatory). To determine the observer accuracy in finding carcasses, a simulation of the searches is made through a known number of carcasses or models that simulate different birds and bats species

from fatality estimators based on mathematical formulas that follow an on-going investigation by the scientific community in recent years (e.g., Huso 2010; Huso et al. 2015; Korner-Nievergelt et al. 2011, 2013); see also Marques et al. (Chap. 7).

Whenever mitigation programmes are proposed during the EIA process, monitoring plans should always include methods that allow the evaluation of their success. This is a procedure that is foreseen in Portuguese guidelines for impact assessment in wind farms and is found also in other international documents (see Appendix, Table 3.5). The evaluation methods to assess the success of mitigation measures must be defined on a case-by-case scenario, depending on the type of mitigation (compensatory and/or minimisation) and the target communities. Nonetheless, the general outcomes are expected to be reflected at population levels (e.g., distribution, abundance, breeding success) when, for example, habitat management measures are implemented (Paula et al. 2011b, 2015a); and/or at specific wind farm local scale (e.g., decrease in mortality rates) when minimisation measures such as curtailment (changes in turbine speed) are implemented (Arnett et al. 2011; Singh et al. 2015). Other mitigation programmes designed to produce changes in the behaviour of target species within wind farms, as in the case of deterrents (e.g., Arnett et al. 2013a) should also be subject to evaluation of its efficiency given the desired goals. The theme on mitigation measures is currently an on-going investigation, and specialists and developers are working to better understand the most effective strategies to implement in wind farms to biodiversity conservations, as may be found in papers on the theme (e.g., Arnett et al. 2013b; Marques et al. 2014; Peste et al. 2015). In addition, for these reasons, if the mitigation scheme reveals not to be effective, the plans should be sufficiently flexible to be adjusted—ideally following an adaptive management approach.

3.2.2.2 Flora and Habitats

The main negative impacts on vegetation (flora and habitats) due to the construction of wind farms include the direct destruction of important habitats (removing the existing native vegetation) and/or destruction of important rare plant species. The proliferation of invasive alien plant species is also a known negative impact (Silva et al. 2011; Passos et al. 2013). Although usually not one of the major subjects in environmental assessment, since, internationally, EIAs tend to consider the negative impacts less significant or even negligible, in comparison with those found for fauna (Silva and Passos 2017), impacts in flora and habitats have a direct relation to biodiversity loss (the ecological processes are strongly tied to the vegetation and the habitats). This biodiversity descriptor is addressed on the *guidelines* developed by the Portuguese Environmental Agency for monitoring plans in wind farms (APA 2010; Appendix—Table 3.5) and has been the subject of studies during construction and operation phases in wind farms (e.g., Bioinsight 2016c). In Portugal, one of the main focuses in recent years is the proliferation of an invasive alien plant species (Passos et al. 2013), since this is a problem that leads to serious changes in ecosystems (Richardson et al. 2000; Vilà et al. 2011) and, according to Marchante et al. (2014), there are currently more than 50 terrestrial alien plant species (trees, shrubs, herbs) described for national territory.

In the early stages of the EIA process, especially during the planning phase of a wind farm, it is crucial that major efforts are made to avoid or minimize these negative impacts. However, some negative actions, such as the direct removal of native vegetation during the construction works are virtually unavoidable due to the terrain interventions necessary to build access roads and the deployment of turbines and transmission lines (Fraga et al. 2008). In the direct implementation areas of the infrastructures, some of the impacts will remain during the lifetime of the wind farm. On the other hand, in the vicinities of the turbines and other support areas of a wind farm, where the terrain intervention is temporary (only during construction phase), the impacts can be significantly reduced. The natural recovery and/or the restoration of the vegetation and associated habitats is possible, but one of the major challenges is to avoid the dissemination of invasive species on site.

The importance of assessing the real impacts on flora and habitats is pointed as a major factor by Silva and Passos (2017). Monitoring programmes during the construction and operation phases of wind farms should include methodological approaches in order to evaluate the following:

- Are the natural plant species recovering in the degraded areas?
- Are there any alien plant species present in the wind farm area or vicinities?
  - if so, are this species spreading in the intervention areas?

Rehabilitation or restoration of disturbed natural vegetation sites is often implemented in projects at the end of the construction phase, aiming to mitigate the negative impacts resulting from the implementation of the wind farm and support better landscape integration. Sampling methods for monitoring and analysing data of natural vegetation recovery or restoration usually follow BACI designs based on key studies (e.g., Braun-Blanquet 1979; Coulloudon et al. 1999) aiming to evaluate population parameters for plant species as growth and species richness (Table 3.4). In the case of invasive alien plant species, preventing, monitoring and eradicating plans to avoid the proliferation of plants are steps that have to work hand-in-hand since the early stages of any project, and increasing efforts are made in Portugal to ensure these strategies (Passos et al. 2013). Specific mitigation plans must be defined in cooperation with developers to ensure only the use of soil from non-invaded areas in recovering areas as well as the correct measures to prevent seed proliferation during the removal of any existing alien invasive species within the wind farm and surrounding areas (Silva and Passos 2017). Mitigation plans should include regular monitoring to control any occurrences (Table 3.4) and proceed to the eradication according to the assessment results and the present species. Different methods may be used—separately or combined—including manually removing of young plants (seedlings) and applying chemical treatments for adult plant destruction (Passos et al. 2013).

3.3 Case Study|The Step Forward in Assessing Biodiversity Impacts

Standard survey protocols to monitor the effects that Portuguese wind farms may have on bird and bat communities generally require low-intensity effort, especially in comparison with other international standards (Table 3.5, see Appendix). As an example, the Scottish guidelines require an average of 12 h per sampling location (SNH 2014), while the South African guidelines for bat monitoring require sampling of bat activity through 100% of the year (Sowler et al. 2016). However, with the increase in impact awareness caused by wind farms, decision-makers are

Table 3.4 Field methods to collect data during wind farm construction and operation

Field method	Target group/Species	Description
Vegetation mapping/Cartography	Natural vegetation/habitats and alien plant species	Transects on the wind farm area. Field surveys for mapping the vegetation cartography in the intervention areas during and after construction works to establish the status of the vegetation cover damage due to the project. Data collection and analysis may include both mapping natural vegetation cover (including important/rare plant species) and updating the mapping of alien plant species ideally accesses before construction (species identification, cluster location and quantity)
Inventory	Natural vegetation	Sample plots—rectangular or quadrats. Field surveys where the observer collects data for several vegetation parameters allowing for the assessment of abundance and distribution of species, identification of the present species and determining ecological status of the habitats. The periodic repetition of the surveys within the same sampling plots leads to a systematic analysis of information to describe, characterise, or quantify vegetation on long-term
Control	Alien plant species	Prospection of the known clusters and new potential dispersal sites of alien species. Control and monitoring programmes include the accurate species identification and specific mapping based on the cluster cartography from the vegetation mapping in previous project phases. The aim is to actively avoid the spread of invasive species; thus, the collected data should be compared periodically. In any case, where new clusters or a change in the known sites occurs (new plants) additional control actions have to be defined to eradicate the alien invasive species, through de-implementation of a new mitigation plan or the reinforcement of any mitigation plan already on course

becoming progressively more demanding of information that is able to determine impact magnitude and significance in the pre-construction phase. In order to provide this information, it is necessary to determine the extent of bird and bat utilisation of the development area where the proposed wind farm is planned.

Thus, new methods have been required to fulfil such requirements while presenting solutions with an adequate cost–benefit ratio. Automated radar systems are one of the solutions able to fill this gap, as they are able to detect and map movements of moving objects. As an example, when flying, birds are seen by radar tracking systems as moving targets and are recognized like airplanes. In comparison with human observers conducting observations of passerine and raptor movements, radar systems are precise and collect data continuously, through all hours of the day, for as long as its deployment lasts. Most systems involve self-working equipment, with a capacity to automatically store all collected data and may be controlled remotely through a standard internet connection.

Though it is not standard practice in Portugal, such methodologies have been implemented in different situations, which are illustrated below.

Example: Radar Utilisation for Bird Detection

The use of Merlin™ Avian Radar System, an adapted marine radar, was used to study bird movements at different landscape scenarios, including mountain landscapes where wind farms are situated and wetlands where birds are abundant. Marine radars are a good option for local scale studies because typically they have a combination of at least two antennas: one vertically scanning X-band 25-kW magnetron radar (VSR) and one horizontally scanning S-band 200-W solid-state Doppler radar (HSR). HSR is particularly useful for characterizing dominant flight directions and target speed, whereas VSR is mainly used to characterise flight altitude. Both variables are very important to determine the characteristics of the bird and bat movements within a development site.

Though radar systems utilisation is particularly advised in relatively plain locations (Fig. 3.3), as rough areas will produce noise and 'black holes' within the sampled area (Fig. 3.4), in Portugal, wind farms are generally placed in areas with accentuated topographic features. In a recent study conducted in Gardunha wind farm, located in a mountainous region, and in 'Lezíria', a plain wetland within Tagus River Estuary Natura 2000 Special Protected Area (both in Central Portugal), Rosa et al. (2016) compared data collected via radar in both sites regarding bird sampling. The experience contributed largely to increasing the knowledge on the utilisation of radar systems at Portuguese wind farms, as well as the main analysis

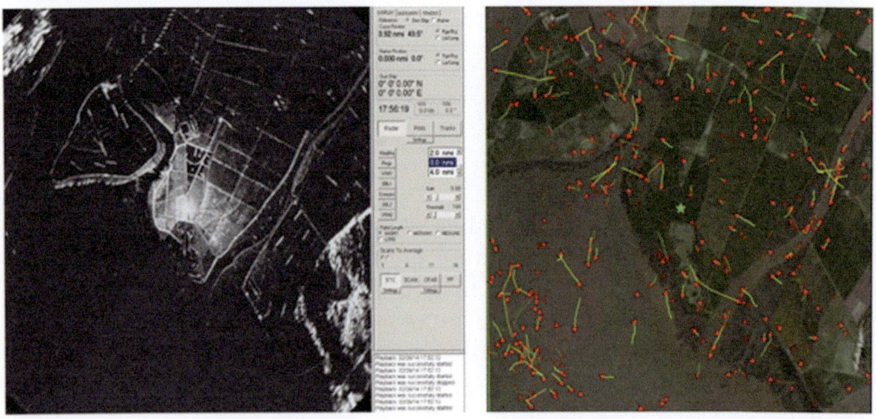

Fig. 3.3 Example of bird movements detected through RADAR system at Leziria, Portugal

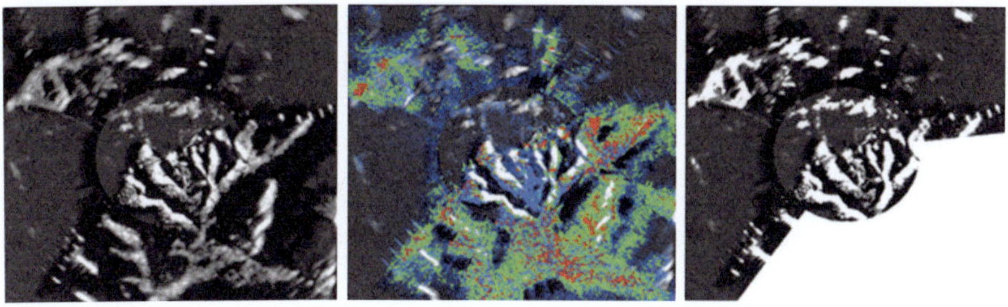

Fig. 3.4 Illustration of the clutter detected at Gardunha wind farm, Portugal, and artificial mask applied to reduce data noise captured during normal deployment

methods that should be used to overcome logistic and terrain difficulties. Thus, when considering the practical applications at a more regular basis, this study allowed better understanding of how radar systems can be applied in EIAs and monitoring programmes in wind farms aimed at bird communities, as well as the main benefits and limitations.

Benefits of the method:

- It is possible to sample the area continuously, including daytime and night-time. This is a significant increase in relation to the sampling of the area using human observers, who can survey the area only during the daytime period;
- Radar systems can survey areas inaccessible to humans, such as really high altitudes without adequate road access, or offshore sites that may be outside of the observation scope for human observers;
- Data collected using radar is not biased by observer characteristics or decrease in detection ability due to fatigue associated with many work hours in human observers. Since radar systems are programmed and collect data in always the same way, this problem is eliminated from avian data analysis;
- The utilisation of radar in optimal areas, within plain areas and with low vegetation allows detailed data to be gathered on bird movements and allows ecological questions to be answered that otherwise would be difficult to determine. As an example, in the study of Rosa et al. (2016), radar use within plain wetlands allowed bird movements associated with disturbance to be determined due to the continuous data gathering.

Limitations detected:

- Utilisation of the radar system in mountainous areas was difficult but not impossible, requiring an increased effort in the deployment phase in order to map the clutter areas and potential black areas (areas that are blocked by mountains and/or buildings). Though these noise situations can be masked, they imply the loss of the observation area, since any movements that happen within those masked areas will not be considered for analysis (it is impossible to separate noise from biological movement in these situations, with the current knowledge);

- Performance of the radar and machine learning techniques' utilisation in separating bird species and/or groups has limitations in: (i) areas with high species diversity or (ii) wherever species/functional groups with similar morphological and behavioural features co-occur;
- Datasets need to include significant information on the species that a researcher is aiming to track as well as for other targets that are potentially being captured by the radar, such as other bird species and non-biological targets, in order to separate them accurately;
- Bat detection is common, since radar systems are used continuously through the night. It is therefore necessary to include bats in the dataset, which becomes an issue because the ground truthing dataset is based on visual identification of the radar targets. Thus, correct identification of bird and bat activity using marine radar data is very complex and might not be a feasible task.

Other RADAR system possible applications for birds: *RADAR as a tool to support mitigation actions*

Due to its characteristics, radar is also useful at wind farms to support the minimisation of actions such as 'turbine shutdown on demand' (e.g., Marques et al. 2014; Birdlife International 2015). This type of action is currently being conducted at several wind farms in southwestern Portugal. These wind farms are located within a migration route for soaring birds, near the cape in Sagres. This is one of the most important coastal areas in Portugal where a high number of birds are concentrated in the migration from the north (Europe) to the south (Africa) in autumn, so the wind energy projects may represent a high risk of mortality due to collision during this specific season. One case study in this area is the Barão S. João wind farm where 'turbine shutdown on demand' actions supported by radar have been taking place since 2010 (Tomé et al. 2012; Birdlife International 2015). The mitigation scheme is based on radar technology and field observations to detect soaring bird flocks, allowing for more efficient detection and the coverage of a larger range, thus providing crucial information to help in the decision process on the need to shutdown turbines (Tomé et al. 2012). The data is used in real-time to temporarily shut down specific turbines when migration movements are high and coincident with turbine blade heights (high collision risk) (GPWind 2013). The results of the mitigation programme suggest a high success on reducing mortality of soaring birds in the Barão S. João wind farm, since for at least the first five years of assessment during the operation phase, no fatalities occurred (STRIX 2015) even though the number of recorded soaring birds involved in movements at the wind energy facility averaged 20,000–30,000 per year (Birdlife International 2015). According to Tomé et al. (2012), a high proportion of turbine shutdowns occur with light winds (<5 m/s), corresponding typically to low electricity production—which in turn represents a low production loss for developers (c.a. 2%; GPWind 2013).

On the other hand, the improvement of target classification techniques can make radar systems more sensitive to specific target species, thus triggering the mitigation action, for example, turbine shutdown, automatically, and only when the desired target is within the vicinity of the wind turbine (Rosa et al. 2016).

Another possible application, as conducted while using the radar in Leziria, would be in agricultural management by detecting when flocks of bird species that are harmful to crops fly to the croplands, and warning farmers to take dissuasive measures.

Example: Radar Utilisation for Bat Detection
Radar utilisation as a methodology for bat detection and monitoring in Portugal is still relatively poorly understood. Hence a study was conducted at the Bird Watching and Visitation Space (EVOA), a wetland within Tagus River Estuary Natura 2000 Special Protected Area, to better understand the potential utilisation of this tool to study and map bat activity using radar systems. A Merlin™ Avian Radar System was used, which was stationed at EVOA installations working from early-June to late-July and from mid-October to early-November of 2013 (Bio3 and Aveiro University 2014). In order to validate the data collected by the radar, and as ground truthing bat activity is very difficult, since it is based on visual identification of radar targets, an automatic ultrasound detector Batcorder 2.0 (installed at EVOA rooftop, at approximately 10 m above ground level and programmed to capture data everyday between 8 p.m. and 8 a.m.) was tested, aiming to validate the targets identified by the radar system.

In order to understand if both data sources exhibit similar variation patterns, information collected from both equipment (considering the same sampling period and area) was compared, based on the number of bat/targets recorded (Bio3 and Aveiro University 2014). However, the data analysis did not show promising results in the utilisation of Batcorder to validate the information collected by radar system, as no significant relation between data collected by the two information sources was observed.

Differences between the two parameters are most likely to be related with bird records captured by the radar (instead of only bat detections), which are probably masking the results obtained for bats, and hindering the identification of similar detection patterns obtained through both equipments. This situation is particularly difficult in locations such as the Tagus River Estuary where bird abundance is very high, even for bird species that conduct movements in low-light conditions (nightfall and before dawn). Thus, the utilisation of automatic ultrasound equipment to validate data collected by radar system has severe limitations to the following:

- For it to be possible to make any correspondence between detections it is required that both equipments are configurated with the exact same time and data between themselves, and no deviations can occur.
- Data comparison between the equipment is not automatic and requires human intervention, thus it is a time-consuming task and demands a high-concentration level;
- Since the equipment for one does not collect the same type of information as the other, there is some uncertainty associated with situations in which one did not present records; for example, when data was collected by the radar but there was no correspondence with the Batcorder, one could assume it was a bird record. However, when the opposite happens, the decision is not simple, and the biological target was likely missed by the radar system.

Example: Bat Automatic Detection Systems to Determine Mitigation Measures
A wind farm was proposed near Évora, in central Portugal, comprising the installation of four wind turbines with a tower height of 85 m and rotor diameter of 90 m. Due to the presence of a nearby roost of national importance, a condition of the authorisation from the Portuguese Environmental Agency was the implementation of specific mitigation measures, which included curtailment of all wind turbines when wind speed was below 5 m/s, between sunset and sunrise, and for all months between March and October, inclusive, through all operational phase of the wind farm. This curtailment value could, in turn, be adjusted after the analysis of a specific study to evaluate bat activity in relation to air temperature and wind speed. This study should take into consideration the following:

- Special attention should be given to all species of the *Nyctalus* genus—which have a higher collision risk compared to other genus, as identified by Marques et al. (Chap. 1)—and also to species with a conservation status of concern (all threatened species by national criteria);
- The wind speed below which bat activity becomes negligible (at this point, a value of 2 bat passes per hour was considered). Other variables that could interact with wind speed effects should also be tested;
- Sampling period should consider the whole year, in such a way that all environmental conditions were captured, and it was mandatory that the study be conducted before wind farm construction.

Given the specifications of the requirements, standard methodologies could not solely be used, since they would not have the capacity to collect enough information and relate this to the environmental variables with the necessary detail. In addition, standard methods are conducted at ground height, when in fact the collision risk area is located where turbine blades are spinning (at rotor height, approximately 50 m height). Hence, an automatic ultrasound detector was used at the met mast installed where the wind farm was proposed in order to sample bat activity continuously and at rotor height (50 m) (Bio3 2014; Fig. 3.5).

Since only one met mast was available at the site and information needed for layout adjustment was necessary (i.e. spatial information of bat activity), active detection of bat activity, with a handheld ultrasound detector was conducted on some sampling nights. This information could be related with bat activity recorded at height since a second automatic detector was placed (at 6 m height) at the met mast, also recording continuously. Thus, the objective was to relate data collected through active detection with information gathered with the automatic detectors. Sampling through automatic detection was conducted between August and October 2013 and March and July 2014 according to the known activity patterns of bats in Portugal (ICNB 2010b). Active detection surveys were conducted in the same time period but in individual surveys spread over this period.

This sampling design allowed the requirements of the competent authority to be answered, by providing indications with regards to the wind speed influence on bat activity where collision risk is higher: this was significantly higher when wind direction was from the south, as well as between 3 and 8 m/s (Bio3 2014).

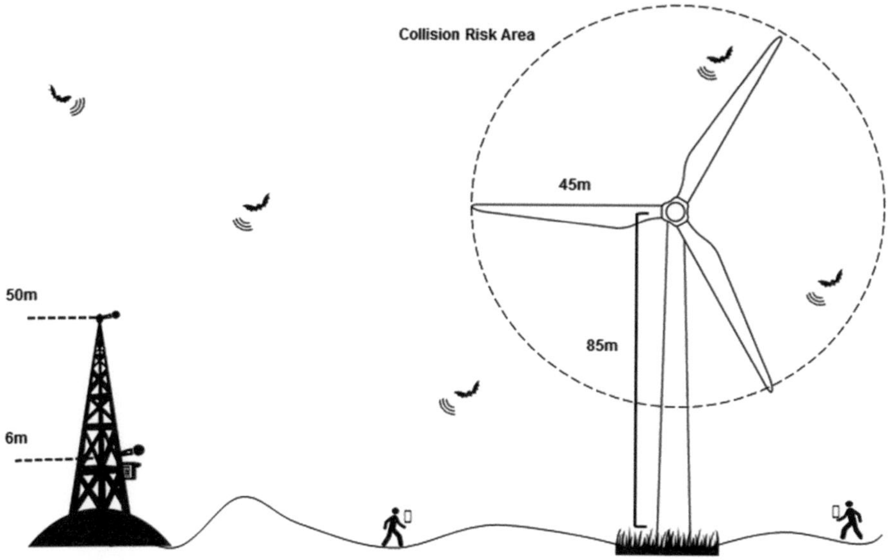

Fig. 3.5 Sampling methods implemented to determine bat activity patterns at a wind farm located near Évora, central Portugal. Adapted from Bio3 (2014)

Although in Portugal this type of study is not common since it is not required by the best practice guidelines, in other parts of the world this is already a requirement since monitoring bat activity at rotor swept area and relating it with wind speed allows adequate and sustained mitigation measures to be proposed (Rodrigues et al. 2015; Sowler et al. 2016).

However, as demonstrated, utilisation of automated bat detectors to spatially cover the area where a wind farm is proposed was challenging, since in a pre-construction phase, no structures with adequate height to sample bat activity had yet been constructed, with the exception of one or two met masts, depending on the proposed area size. Unlike standard monitoring with a handheld detector, where one detector can sample a multitude of locations, the utilisation of an automatic device restricts the number of locations sampled, since detectors are left at one location, for a long period of time, especially due to the logistics involved in the placement of a microphone at rotor height (usually involving one climbing team with special skills to conduct this type of work). Spatial modelling techniques (e.g., MaxEnt) may allow this problem to be overcome by indicating features of high bat activity in the general area, using data collected at height as the sampling unit (Philips and Dudík 2008).

3.4 Final Remarks

To date, the combined work of sector's developers, environmental authorities, specialised consultants and investigators has provided significant knowledge both on the impacts from wind farms installations that may affect biodiversity and also

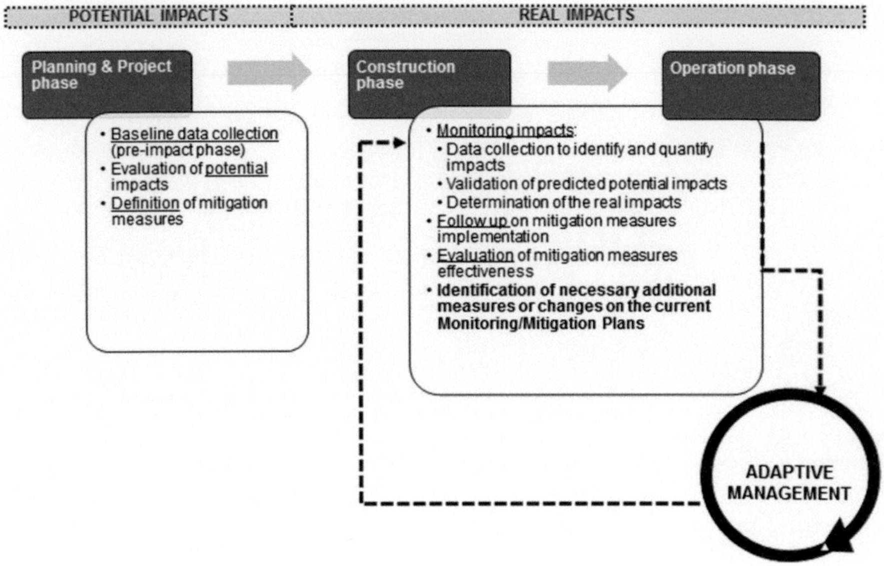

Fig. 3.6 Summary scheme of the Environmental Impact Assessment process: from pre-construction to operation phase in a wind farm. Adapted from Marques et al. (2015)

on the main methods to assess them. Although there are some differences in the approaches on impact assessment found across countries, the main strategy remains the same and Portugal is no exception (Fig. 3.6): to understand the interactions between wind farms and biodiversity as early as possible in the stages of project developments and define the necessary means to mitigate the potential impacts; assess the real impacts and evaluate the effectiveness of any mitigation measures and monitoring plans and review them whenever necessary.

There are, however, some questions that require further investigation. One example is the challenge that still remains on estimating the real extent of the effects of wind farm driven mortality over wildlife at the population levels. This is still a difficult matter since robust information of species population size and boundaries (specially migrating species) is still insufficient. In Portugal, for example, this is a question that is typically not addressed in monitoring reports. Internationally, methods are currently being researched in order to estimate the real mortality rates and the impact of these rates on species conservation, so it is expected that Portugal quickly follows this trend.

Appendix

(See Table 3.5).

Table 3.5 Compilation of the analysed guidelines for impact assessment (pre- to post-construction phases) in wind energy projects developed for Portugal in comparison with other international references

Country and Bib. Ref.	Target groups	Document purpose	Obligation	EIA phase	Assessment goals
Portugal (APA 2010)	Large carnivores (Iberian wolf)	Guidelines for the environmental impact assessment in wind farms	Best practice guidelines (not legal requirement) After being considered at a scoping level, it becomes mandatory to be included in the Final Impact Assessment and following stages of the process	Specialist Impact Assessment Studies Environmental Impact Assessment Report Pre- and Post-Construction monitoring	BACI approach on the assessment of the main Impacts: (a) Displacement/avoidance (b) Barrier effects (c) Reproductive fitness (if long term data is available) (d) Cumulative impacts (e) Effectiveness of mitigation measures/validation on the requirement for additional mitigation measures
	Birds				BACI approach on the assessment of the main Impacts: (a) Displacement/avoidance (b) Mortality (c) Effectiveness of mitigation measures/validation on the requirement for additional mitigation measures
	Bats				BACI approach on the assessment of the main Impacts: (a) Displacement/avoidance (b) Mortality

(continued)

Table 3.5 (continued)

Country and Bib. Ref.	Target groups	Document purpose	Obligation	EIA phase	Assessment goals
					(c) Effectiveness of mitigation measures/validation on the requirement for additional mitigation measures
	Flora and Habitats				BACI approach on the assessment of the main Impacts: (a) Habitat destruction (b) Invasive species proliferation (c) Effectiveness of mitigation measures/validation on the requirement for additional mitigation measures
Portugal (ICNB 2010b)	Iberian wolf	Orientations for impacts of infrastructures on Iberian wolf	Best practice guidelines	Specialist Impact Assessment Studies Environmental Impact Assessment Report Pre- and Post-Construction monitoring	BACI approach on the assessment of the main impacts: (a) Displacement/avoidance (b) Barrier effects (c) Reproductive fitness (if long term data is available) (d) Cumulative impacts (e) Effectiveness of mitigation measures/validation on the requirement for additional mitigation measures

(continued)

Table 3.5 (continued)

Country and Bib. Ref.	Target groups	Document purpose	Obligation	EIA phase	Assessment goals
Portugal (ICNB 2009)	Bats	Recommendations on bat monitoring in wind farms	Best practice guidelines	Pre- and Post-Construction monitoring	BACI approach: methods should allow identification of the following major effects: (a) Displacement/avoidance (b) Mortality
Spain (Atienza et al. 2011)	Birds and Bats	Guidelines to assess the impact of wind farms on birds and bats	General methodological guidance	Strategic Environmental Assessment	BACI approach: methods should allow identification of the following major effects: (a) Displacement/avoidance (b) Barrier effects (c) Collision mortality
Scotland (SNH 2009a, b)	Birds	Definition of standard methodologies to be adopted in monitoring birds at onshore wind farms: detailed advice on the design and methodology	Defined as best practice and should be adopted at all monitoring programmes for bird in WF locations	Monitoring according BACI design: – Pre-construction and operation	Methods should allow to identify the following major effects: (a) Displacement and disturbance of breeding and/or non-breeding birds (b) Collision mortality (with turbines, buildings and overhead power lines) (c) The barrier effect to dispersing and/or migrating birds (d) Habitat loss (direct) from the wind farm infrastructure (long-term habitat change also mentioned)

(continued)

Table 3.5 (continued)

Country and Bib. Ref.	Target groups	Document purpose	Obligation	EIA phase	Assessment goals
Scotland (SNH 2014)	Birds	Defines a set of standards for bird surveys at proposed wind farms on land: (1) Pre-survey requirements needed to define a site-specific survey programme (2) Detailed recommendations for survey methods	Best practice: based on EIA regulations in Scotland	Pre-construction	Survey is usually required to: (a) Inform turbine layout, species protection plans (particularly with specially protected species), and mitigation including habitat management plans (b) Forms a baseline for any future monitoring
United Kingdom and Ireland (CIEEM 2016)	Terrestrial, freshwater and coastal	General guidance for adequate ecological environmental assessments Not specific for wind energy developments	Mandatory	Scoping impact assessment Pre-construction and final impact assessment	Establish baseline for ecological impact assessment
EUROPE (Rodrigues et al. 2015)	Bats	Guidelines for consideration of impacts of wind energy on European bat populations	Best practice guidelines	Impact Assessment Pre- and Post-Construction monitoring	Pre-construction survey is usually required to: (a) Identify habitats and areas use by bats (b) Form a baseline for any future monitoring Monitoring methods should allow to identify the following major effects: (a) Area use by bats and breeding success (b) Displacement and avoidance (c) Collision mortality

(continued)

Table 3.5 (continued)

Country and Bib. Ref.	Target groups	Document purpose	Obligation	EIA phase	Assessment goals
USA (Strickland et al. 2011)	Birds and Bats	Guidance document to persons involved in designing, conducting, or requiring wind energy/wildlife interaction studies	Best practice guidelines The document provides a reference for use in the production of a body of scientific Information adequate to fulfil the assessment goals	Impact Assessment (pre-construction) Post-Construction monitoring	Site selection/Pre-construction survey is usually required to: (a) Assess the suitability of a proposed wind facility site with regard to species of concern, including the potential for fatalities and for habitat loss Monitoring methods (BACI approach) should allow identification of the following major effects (potential and real) (1) Displacement and avoidance (2) Breeding success (3) Collision risk (4) Mortality rates by collision Should also be evaluated the effectiveness of measures taken to avoid, minimise, or offset significant adverse impacts and risk reduction management actions to reduce future impacts

(continued)

Table 3.5 (continued)

Country and Bib. Ref.	Target groups	Document purpose	Obligation	EIA phase	Assessment goals
South Africa (Sowler et al. 2016)	Bats	Guidelines for evaluation a of impacts of wind energy on bat populations: pre-construction phase	Best practice guidelines	Scoping Impact Assessment Bat specialist impact assessment report: Pre-construction monitoring	Survey is usually required to: (a) Assess the suitability of a proposed wind facility site with regard to species of concern, including the potential for fatalities and for habitat loss (b) Assess the potential effects of a proposed wind energy facility on species of concern such as: (1) Displacement and avoidance (2) Breeding success
South Africa (Aronson et al. 2014)	Bats	Guidelines for assessment of impacts of wind energy on bat populations: operational phase	Best practice guidelines	Post-construction monitoring	Methods should allow to: Validate of predicted impacts Determinate of real impacts such as (a) Displacement and avoidance (b) Bat fatalities
South Africa (Jenkins et al. 2015)	Birds	Best practice guidelines for assessing and monitoring the impact of wind-energy facilities on birds in Southern Africa	Best practice guidelines	Scoping Assessment Bird Specialist Impact Assessment: Pre-construction monitoring	Scoping: To define scope of EIA assessment and characterise the study area and the broader area of influence Multi-tiered approach, with the aims of:

(continued)

Table 3.5 (continued)

Country and Bib. Ref.	Target groups	Document purpose	Obligation	EIA phase	Assessment goals
				Construction monitoring Operational monitoring	(a) Informing current environmental impact assessment processes (b) Developing our understanding of the effects of wind energy on southern African birds; and (c) Identifying the most effective means to mitigate these impacts Pre-construction monitoring and avifaunal impact assessment: informs the impact assessment and decision-making process; provides a baseline against which post-construction monitoring can be compared Construction-phase monitoring: may be necessary Post-construction monitoring: document impacts and identify if additional mitigation is required – Detailed research to address specific questions, may be required where there is a need to understand and mitigate impacts

(continued)

Table 3.5 continued

Methods	Duration (WF phase)
Displacement, disturbance and avoidance: *Study area*: Proposed development area (5 km radius around WF) and Control area *Surveys scale*: based on a 2 km × 2 km grid with at least two sampling locations within each grid cell – Methodology: to be defined with relevant authorities (may include camera traps, transects, scat surveys, etc.) Different teams in the same regional area should adopt similar methodologies contributing to cumulative impact assessment	1. Pre-construction: 1 year 2. Construction: full coverage 3. Operation: minimum 5 years Monitoring plans to be reviewed annually and adjusted (if required)
Displacement, disturbance and avoidance: *Study area*: Wind farm construction area and control area – Population surveys covering a full annual cycle (four seasons: wintering communities, breeding, juvenile dispersion and migration) – Two sampling surveys per season – Passerines: count points in 250 m radius – Large raptors and soaring birds: *not specified* – For species particularly sensitive to wind energy impacts – Target species to be defined and specific assessment methodologies to be defined and implemented *Fatality assessment*: – Searches should be made on all months of the year: weekly in one or two months of each wintering, breeding, juvenile dispersion and autumn migration seasons; and monthly in the remaining months – Searches must be conducted in all WF or at least in 70% of them It is recommended that the duration of search will be 20 min in a 50 m radius around the WF – Correction factors such as carcass removal rate and searcher detection rate should be calculated to minimise the bias associated with searches and obtain a fatality estimation. Thus, trials must be carried out in same seasons as the search	1. Pre-construction: 1 year 2. Construction: full coverage 3. Operation: minimum 3 years Monitoring plans to be reviewed annually and adjusted (if required)

(continued)

Table 3.5 (continued)

Methods	Duration (WF phase)
Displacement, disturbance and avoidance: *Study area*: Wind farm construction area and Control area – Bat activity assessment: Point counts (10 min sampling points) – Usage of the study area/roosts (within 10 km radius): Individual counts in known roosting sites and species utilisation (hibernation and reproduction) – For species particularly sensitive to wind energy impacts: – Target species to be defined and specific assessment methodologies to be defined and implemented *Fatality Assessment*: – Carcass searches during operational phase – Weekly from March to October – Searches must be conducted in all WFs or at least in 70% of them – Correction factors such as carcass removal rate and searcher detection rate should be calculated to minimise the bias associated with searches and obtain a fatality estimation	1. Pre-construction: 1 year 2. Construction: full coverage 3. Operation: minimum 3 years Monitoring plans to be reviewed annually and adjusted (if required)
Monitoring requirements should be defined case by case if sensitive species or Habitats are identified in the project area (list in Law Decree, based in the Habitats Directive 92/43/CEE) *Study Area*: wind farms, details not specified *Surveys*: – Pre-construction phase—flora and habitat cartography – Operation phase—review cartography (vegetation loss; recovery; invasive species spread); *parameters not defined* If necessary, management actions should include an invasive species control plan	1. Pre-construction: recommended but not specified 2. Construction phase: not specified 3. Operation: recommended but not specified Monitoring plans to be reviewed annually and adjusted (if required)

(continued)

Table 3.5 (continued)

Methods	Duration (WF phase)
Displacement, disturbance and avoidance: *Study area:* Proposed development area (5 km radius around WF) and control area *Surveys scale*: based on a 2 km × 2 km to 5 km × 5 km grid with at least two sampling locations within each grid cell; In each grid, two constant sampling points must be chosen and some mobile sampling points depending on previous results – Methodology: to be defined with relevant authorities (may include camera traps, transects, scat surveys, telemetry, etc.) Depending on the method used, periodicity may vary. Transects must be done monthly, camera trapping should be done throughout the whole year, never less than 12 days/month	1. Pre-construction: 1 year 2. Construction: full coverage 3. Operation: minimum 5 years Monitoring plans to be reviewed annually and adjusted (if required)
Displacement, disturbance and avoidance: *Study area*: Wind farm construction area and control area – Bat activity: Point counts (10 min sampling points) – Specifies weather conditions and period of day/night to proceed to sampling data; survey period (March–October); and minimal distance between sampling points – Reproductive fitness: roost search (10 km radius around proposed WF site) – Specifies minimum survey period: if roosts have more than 20 animals inside or guano on the floor, they must be visited once a year *Fatality Assessment*: – Carcass searches during operation phase – Linear transects in a minimum of 50 m radius around each turbine – Adjustment factors should be calculated, such as carcass removal rate and searcher detection rate to minimise the bias associated with searches and obtain a fatality rate as close as possible to the real figure	1. Pre-construction: 1 year 2. Construction: recommended but no specifications are made 3. Operation: 3 years After the first three years, the results must be evaluated for each case and continuation of monitoring decided or not

(continued)

Table 3.5 (continued)

Methods	Duration (WF phase)
Displacement, disturbance and avoidance: *Study area*: Wind farm construction area and Control area – *Birds*—sampling methods: – Survey: fixed width line transects 500 m length × 25 m width (density estimation); point counts (5–10 min × 30–40 m radius) if transects are not possible (e.g., depending on habitat or present species) – Other suggestions are specified if there are target species/communities where these methods are not appropriate (e.g., raptors) – Sampling period: first hours in the morning; sampling frequency throughout the year is not specified – Document states that pre-construction methodology (EIA report) should be suitable for repetition during operation phase – *Bats*: during pre-construction, the EIA report recommends following the suggestions described in EUROBATS (see Rodrigues et al. 2015). No specifications are made for impact assessment during operation phase: one can assume this to be the same as for pre-construction *Fatality Assessment*: – Search area must be adapted to specific characteristics of terrain and vegetation; specifications are presented in the document – Carcass searches by expert observers (specifies search effort number of turbines/day) – Sampling effort: searches (in every turbine) every 15 days in the first 3 years; monthly from the fourth year onwards (number of turbines to be checked decided based on the size and years on operation of the WF) – Adjustment factors should be calculated, such as carcass removal rate and searcher detection rate to correct the fatality estimates – Searches outside influence zone (~500 m) of the monitored WF should be carried out to find out the background fatality rate of the study zone	1. <u>Pre-construction</u>: recommended but no specifications are made 2. <u>Construction</u>: not specified/not addressed 3. <u>Operation phase</u>: at least 3 years

(continued)

Table 3.5 (continued)

Methods	Duration (WF phase)
Displacement, disturbance and avoidance *Study area*: Wind farm construction area and control (reference) site Monitoring of breeding and non-breeding birds (migrants) – Specified survey methods for each bird community (according to ecology, phenology, behaviour, reproduction, etc.) – Field methods: mainly area-based counts; alternate methods include point counts; line transects (distance sampling); vantage points; also refers to species where specific methods have to be applied (colonial species; breeding raptors, etc.) – Sampling period: is well defined to each group, phenology and/or survey method Collision risk studies and vantage point watches – Detailed description of the method: main targets are diurnal species of conservation importance – Quantify flight activity in wind farm area to determine collision risk and behaviour changes due to infrastructures – Survey period: spread over the whole season according to target species (breeding, passage or wintering)—recommended time of sampling: at least 36 h per season *Fatality Assessment*: Survey methods by human observers: *not clearly specified* (field methods, search period and frequency not specified). Correction factors to mortality estimates: – *Search bias*: an approach to a calculation method for the proportion of carcasses falling outside the search area is described – *Observer detection bias*: recommendations are given to observer detection trials. Trial methods are *not clearly specified* – *Scavenger bias*: recommendations to carcass removal experiments. Trial methods are *not clearly specified*	1. Pre-construction: don't have a clear definition (defined in SNH 2014) 2. Construction: recommended but not specified 3. Operation: (a) Standard recommends that monitoring takes place over at least 15 years (periodically), after the wind farm becomes operational: years 1, 2, 3, 5, 10 and 15 (b) Special case in projects Where major habitat change has taken place—every 3 years: years 3, 6, 9, 12 and 15 after commissioning

(continued)

Table 3.5 (continued)

Methods	Duration (WF phase)
Essential pre-survey requirements needed to define a site-specific survey programme: (a) According to three main potential risks to birds: habitat loss; displacement; collision (b) Special attention to designated sites for birds in Scotland (c) Count on experiment observers/specialists (d) Requires assessment of cumulative impacts (SNH 2012 provides a detailed guidance on this matter) (e) Recommendations for determining the level of survey work (EIA assessment depth) – According to the scale of the wind farm (planned size and location) and bird interest in the area: this factors will define target species and duration of survey period (f) Desktop work to partially assess points (b) to (e) Detailed recommendations for survey methods At this stage, the need for a BACI approach should be defined (establish reference data for operation phase monitoring, if required, according to projected WF specifications) (a) *Distribution and abundance surveys*: record numbers and distribution of breeding, wintering and migrant birds using the site (b) *Vantage point surveys*: watches from a fixed location to quantify the flight activity of birds at a proposed development site data to estimate likely collision risk (c) For both (a) and (b) detailed survey methods are given—similar to those found in SNH (2009b) (see above)	Pre-construction: monitoring at least for 2 years to establish a reference situation —surveys should cover all time of the year when the target species are present
Specific methods for assessment of wind energy developments are not provided	Not addressed

(continued)

Table 3.5 (continued)

Methods	Duration (WF phase)
Pre-construction assessment: habitat use: – Recommends review of information sources (specified in the document) to identify potential habitats for bats (breeding/roosting and migrating populations) at the location and its vicinities *Displacement, disturbance and avoidance*: *Study area*: Proposed development area (document does not address the need for a control area) – Bat activity in WF: point counts, including (i) manual bat detector surveys on ground level throughout the bat active season; (ii) automated bat detector surveys to be carried out during each manual detector survey visit – General area use and breeding success: – Investigation of important roost sites: new roosts should be searched in a 2 km radius and known roosts should be checked over at least a 5 km radius to assess stages of activity and behaviour of local populations; and also dispersal of colonies and the start of autumn migration. Potential important roosts (including maternity and hibernation roosts as a minimum) should be subject to detailed survey – Sampling period: between May and October *Fatality assessment*: – Plot size radius should be equal to total height of the WF – Number of wind turbines sampled should be, if possible 100% of the WF, if not, a subsample may be randomly selected, stratified by habitat and/or WF features – Searches should walk each transect at a slow and regular place, looking for fatalities in both sides of the line – Search surveys must be conducted in a smaller time interval as possible. It is recommended at least a visit every 3 days – The entire activity cycle of bats should be assessed	1. Pre-construction: recommended but not specified 2. Construction: recommended but no specifications are made 3. Operation: at least 3 years

(continued)

Table 3.5 (continued)

Methods	Duration (WF phase)
– Adjustment factors should be calculated such as carcass removal rate and searcher detection rate to minimise the bias associated with searches and obtain a fatality rate as close as possible to the real figure	
Site Evaluation and Impact assessment – Recommends computerised mapping tool when relevant wildlife data is available along with protected areas to understand if the wind development may or may not result in significant adverse impacts in wildlife – Recommends review of information sources (specified in the document) to identify and mapping vegetation and wildlife habitat relationships in order to the facility developer can decide whether to continue the process – Site visit is referred, but not specified *Displacement, disturbance and avoidance* *Study area*: Wind farm construction area and control (reference) site – *Birds* (activity)—sampling methods: – Survey: point counts (diurnal and some nocturnal species); transect surveys (same objective as points, but more suitable for open habitats) – Other suggestions are specified if there are target species/communities where these methods are not appropriate (e.g. hawks) – *Birds* (breeding success)—recommended sampling methods include raptor nest surveys; radio telemetry – *Bats* (activity and roosting)—sampling methods: – Surveys: acoustic Monitoring (echolocating bats) – Refers also other study methods applicable to bats, depending on target species or assessment goals: Mist-Netting, Exit Counts, Roost Searches, Radar	1. Pre-construction: not specified/not addressed 2. Construction: not specified/not clear (from less or 1 year to several years) 3. Operation: – Species activity: at least 5 years, but may not be continuous—surveys on a less-than-annual basis, e.g., 1, 3, and 5 years after construction – Fatalities: in general, assumes at least 2 years of data collection (exceptions can be made to reduce monitoring duration depending on the existing information near the area)

(continued)

Table 3.5 (continued)

Methods	Duration (WF phase)
Collision risk: – Birds: specifications for modelling of collision risk are given *Fatality assessment*: – Recommended that at least 30% of WFs must be searched, and a minimum of 10 wind turbines, through linear transects – Search interval of 7 days in most cases to answer post-construction fatality questions – Search plots for birds have approximately the radius of the maximum distance from the ground to the highest point on the rotor swept area (~90–120 m) – Correction factors such as carcass removal rate and searcher detection rate should be calculated (periodically) to minimise the bias associated with searches and obtain a fatality estimation	
Displacement, disturbance and avoidance (potential impact assessment): – *Activity surveys*: – Manual surveys: walked or driven transects or sampling with acoustic monitoring – *Passive monitoring*: – Continuous recording ultrasonic bat detectors (minimum 1 microphone installed at rotor height per 10,000 ha developable area) Bat sampling stations at ground level to be determined according to available relevant bat micro-habitats – *Other methodologies (not mandatory, case specific dependent)*: – Mist netting (especially for fruit bats) – *Breeding success* (*roost surveys*): – Daytime inspections in summer and winter; non-invasive roost surveys and commuting routes survey	Brief reconnaissance site visit during Scoping Pre-construction: 1 year (minimum) – Specifications for *breeding success*: 75% total nights of the year. Aiming 100% for reproduction and breeding season

(continued)

Table 3.5 (continued)

Methods	Duration (WF phase)
Displacement and avoidance: Same protocols implemented during pre-construction (Sowler et al. 2016) *Fatality assessment*: – Carcass searches surveys—1 week intervals in first year – Bias factors determination (searcher efficiency and carcass persistency)	Post-construction: minimum 2 years Also refers to auditing for impacts during life of development
Desktop study: – Define priority/target species – Provide initial estimation of likely impacts of the wind development Determine the nature and scope of pre-construction monitoring required Bird species richness and passage rates should also be monitored at a reference (control) site to help understand the causes of any changes observed (i.e. environmental conditions vs. impacts of the WEF itself) *Bird density*: Walked transects, point counts and checklist methods Number of large terrestrial birds: Census counts, road counts Focal point surveys/monitoring: wetlands, roosting and/or breeding sites, met masts, infrastructures should be monitored *Bird movements:* vantage point surveys (minimum 12 h per point per season. Minimum 75% developable area to be covered) Use of radar and other methods (radio tracking, PPT, etc.) (recommended for specific conditions)	Brief recognaissance site visit during Scoping 1. Pre-Construction: minimum 12 months' period, including sampling representative of full spectrum of environmental conditions likely during annual cycle 2. Construction: to be defined at EIA phase 3. Post-construction: minimum first 2 years of operation, then again at the 5th year and every 5 years thereafter

(continued)

Table 3.5 (continued)

Methods	Duration (WF phase)
May not be necessary for all WFs but could be recommended by the specialist in the impact assessment if there is a focal site of specific interest or concern, and/or if there is a need to gather additional data on a species potentially affected by the WEF *Fatality assessment*: Bird carcass searches and bias factors determination (searcher efficiency and carcass persistency) Construction monitoring: to be defined during pre-construction and EIA phase Post-construction: same protocols implemented during pre-construction – Detailed research (if required): determined on a case-by-case scenario	

References

ABO UK. (2014). *Green Burn Wind Farm—Scoping Report*. ABO UK Ltd. Available from http://www.abo-wind.com/uk/images/PDF/Green-Burn-Scoping-Report.pdf

APA. (2010). Guia Para Avaliação De Impacte Ambiental em Parques Eólicos [Methodological Guide to the Environmental Impact Assessment for Wind Farms]. Agência Portuguesa do Ambiente [Portuguese Environmental Agency]. Available from http://www.apambiente.pt/ (in Portuguese).

APREN. (2016). Produção de eletricidade por fonte em 2015 [Electricity production by source in 2015]. Associação Portuguesa de Energias Renováveis—APREN [Portuguese Renewable Energy Association]. Available from http://www.apren.pt/ (in Portuguese).

Arnett, E. B. (2006). A preliminary evaluation on the use of dogs to recover bat fatalities at wind energy facilities. *Wildlife Society Bulletin, 34*(5), 1440–1445.

Arnett, E. B., & Baerwald, E. F. (2013). Impacts of wind energy development on bats: Implications for conservation. *Bat Evolution, Ecology, and Conservation* (pp. 435–456). New York: Springer.

Arnett, E. B., Baerwald, E. F., Mathews, F., Rodrigues, L., Rodríguez-Durán, A., Rydell, J., et al. (2016). Impacts of wind energy development on bats: A global perspective. Chapter 11. In C. C. Voigt & T. Kingstom (Eds.), *Bats in the anthropocene: Conservation of bats in a changing world*. Berlin: Springer.

Arnett, E. B., Hein, C. D., Schirmacher, M., Huso, M. M. P., & Szewczak, J. M. (2013a). Evaluating the effectiveness of an ultrasonic acoustic deterrent for reducing bat fatalities at wind turbines. *PLoS ONE, 8*(6), e65794. doi:10.1371/journal.pone.0065794.

Arnett, E., Huso, M. M. P., Schirmacher, M. R., & Hayes, J. P. (2011). Altering turbine speed reduces bat mortality at wind-energy facilities. *Frontiers in Ecology and the Environment, 9*(4), 209–214.

Arnett, E. B., Inkley, D. B., Johnson, D. H., Larkin, R. P., Manes, S., Manville, A. M., et al. (2007). *Impacts of wind energy facilities on wildlife and wildlife habitat*. Maryland: Bethesda.

Arnett, E. B., Johnson, G. D., Erickson, W. P., & Hein, C. D. (2013b). *A synthesis of operational mitigation studies to reduce bat fatalities at wind energy facilities in North America*. A report submitted to the National Renewable Energy Laboratory. Bat Conservation International. Austin, Texas, USA.

Aronson, J., Aiken, S., Hein, C., Jacobs, D., MacEwan, K., Marais, W., et al. (2014). *South African good practice guidelines for operational monitoring for bats at wind energy facilities*.

Atienza, J. C., Martín Fierro, I., Infante, O., Valls, J., & Domínguez, J. (2011). Directrices para la evaluación del impacto de los parques eólicos en aves y murciélagos (versión 3.0). SEO/BirdLife, Madrid (translated into English as Guidelines for Assessing the Impact of Wind Farms on Birds and Bats (Version 4.0)).

AWWI. (2015). *Wind turbine interactions with wildlife and their habitats: A summary of research results and priority questions*. Washington, DC. Available from https://awwi.org/

Barclay, R. M. R., Baerwald, E. F., & Gruver, J. C. (2007). Variation in bat and bird fatalities at wind energy facilities: Assessing the effects of rotor size and tower height. *Canadian Journal of Zoology, 85*, 381–387.

Barré, K., Leviol, I., Julliard, R., & Kerbiriou, C. (2015). *How offset residual impacts on bats and birds of wind farm?* Poster. France. Available from https://colloque.inra.fr/dim-astrea-jdd2015/content/download/3392/34797/version/1/file/poster_astrea_Kevin_Barre_2015.pdf

Barrios, L., & Rodríguez, A. (2004). Behavioural and environmental correlates of soaring-bird mortality at on-shore wind turbines. *Journal of Applied Ecology, 41*, 72–81.

Bastos, R., Pinhaços, A., Santos, M., Fernandes, R. F., Vicente, J. R., Morinha, F., et al. (2016). Evaluating the regional cumulative impact of wind farms on birds: How can spatially explicit dynamic modelling improve impact assessments and monitoring? *Journal of Applied Ecology, 53*, 1330–1340.

BBOP. (2012). *Standard on biodiversity offsets*. Business and Biodiversity Offsets Programme, BBOP, Washington, D.C. Available from http://bbop.forest-trends.org/

BBOP. (2013). *To No Net Loss and Beyond: An Overview of the Business and Biodiversity Offsets Programme (BBOP)*. Business and Biodiversity Offsets Programme. Washington, D.C. Available from http://www.forest-trends.org/publication_details.php?publicationID=3319

Bernardino, J., Bispo, R., Mascarenhas, M., & Costa, H. (2012). Are we properly assessing bird and bat mortality at onshore wind farms? In *Proceedings of the 32nd Annual Meeting of International Association for Impact Assessment*. Porto, Portugal, May 27–June 1, 2012.

Bernstein, M. A., Griffin, J., & Lempert, R. (2006). *Impacts on energy expenditures of use*. Technical report prepared for the energy future coalition, RAND Corporation, Santa Monica, CL.

Bibby, C. J., Burgess, N. D., Hill, D. A., & Mustoe, S. (2000). *Bird census techniques* (2nd ed.). London: Academic Press.

Bio3. (2005). Estudo de Impacte Ambiental do Parque Eólico de Chão Falcão II [Environmental Impact Study for Chão Falcão II Wind Farm] (in Portuguese).

Bio3. (2011). Monitorização Da Comunidade de Aves no Parque Eólico de Prados – Relatório I (Anterior à Construção) [Monitoring Bird Communities at Prados Wind Farm—Report I (Pre-construction)]. Relatório para ENEOP2. Bio3, Almada Portugal (in Portuguese).

Bio3. (2012). Monitorização Da Comunidade de Aves no Sub-Parque Eólico de Mértola – Relatório I (Anterior à Construção) [Monitoring Bird Communities at Mértola Wind Farm—Report I (Pre-construction)]. Relatório para ENEOP2. Bio3, Almada Portugal (in Portuguese).

Bio3. (2014). Estudo de avaliação da atividade de quirópteros no Parque Eólico de Santiago do Escoural. Relatório final (agosto – outubro 2013 e março – julho 2014) [Monitoring of Bat Communities at Santiago do Escoural Wind Farm—Final Report]. Relatório elaborado para PERFORM3, Parques Eólicos, Lda. Bio3, Lda. Almada, setembro de 2014 (in Portuguese).

Bio3, & Aveiro University. (2014). Estudo da atividade e espécies de morcegos presentes no EVOA (Estuário do Tejo). T4.2 Estudo da dinâmica de aves e morcegos em parques eólicos nacionais. Wind and Biodiversity—Projeto 11541. Co-funded "Mais Centro", "QREN" and European Union (in Portuguese).

Bioinsight. (2016a). Monitorização da Comunidade de Aves e Quirópteros no Parque Eólico de Prados - Relatório Final (Fase de Exploração) [Monitoring Bird and Bat Communities at Prados Wind Farm—Final Report]. Relatório para TrustWind. Bioinsight, Odivelas Portugal (in Portuguese).

Bioinsight. (2016b). Monitorização da Comunidade de Aves e Quirópteros no Parque Eólico de Malhanito - Relatório V (Fase de Exploração) [Monitoring Bird and Bat Communities at Malhanito Wind Farm—Report V (Operation phase)]. Relatório para ENEOP2, S.A. Bioinsight, Odivelas Portugal (in Portuguese).

Bioinsight. (2016c). Monitorização da Flora e Vegetação Parque Eólico de Malhanito - Relatório V (Fase de Exploração) [Monitoring Flora communities at Malhanito Wind Farm—Report V (Operation phase)]. Relatório para ENEOP2, S.A. Bioinsight, Odivelas Portugal (in Portuguese).

Birdlife International. (2015). *Review and guidance on use of "shutdown-on-demand" for wind turbines to conserve migrating soaring birds in the Rift Valley/Red Sea Flyway Facility*. Amman, Jordan. Available from http://migratorysoaringbirds.undp.birdlife.org/

Blanco, J. C., & González, J. L. (Eds.). (1992). *Libro rojo de los Vertebrados de España* (714 pp). ICONA, Colección Técnica.

Braun-Blanquet, J. (1979). *Fitosociologia. Base para el estudio de las comunidades vegetales*. Ed. Blume. Madrid.

Britzke, E. R., Gillam, E. H., & Murray, K. L. (2013). Current state of understanding of ultrasonic detectors for the study of bat ecology. *Acta Theriologica, 58,* 109–117.

CIEEM. (2016). *Guidelines for ecological impact assessment in the UK and Ireland: Terrestrial, freshwater and coastal* (2nd ed.). Winchester: Chartered Institute of Ecology and Environment Management.

Coulloudon, B., et al. (1999). *Sampling vegetation attributes, technical reference 1734-4*. Bureau of Land Management. Denver, Colorado. BLM/RS/ST-96/002+1730. [online] URL: http://www.blm.gov/nstc/library/pdf/samplveg.pdf

de Lucas, M., Janss, G. F. E., & Ferrer, M. (Eds.). (2007). *Birds and wind farms: Risk assessment and mitigation* (p. 275). Madrid: Quercus. ISBN 978-84-87610-18-9.

DGEG. (2016). Energia em Portugal – 2014 [Energy in Portugal—2014]. Direção-Geral de Energia e Geologia [Directorate-General for Energy and Geology of the Ministry of Economy]. Available from http://www.apren.pt/fotos/newsletter/conteudos/energia_em_portugal_2014_dgeg_1459441498.pdf (in Portuguese).

Drewitt, A. L., & Langston, R. H. W. (2006). Assessing the impacts of wind farms on birds. *Ibis, 148,* 29–42.

Drewitt, A. L., & Langston, R. H. W. (2008). Collision effects of wind-power generators and other obstacles on birds. *Annals of the New York Academy of Sciences, 1134*, 233–266.

E2P—Energias Endógenas de Portugal. (2015). Parques Eólicos em Portugal – December 2015 [Wind Farms in Portugal]. INEGI—Institute of Mechanical Engineering and Industrial Management and APREN [Portuguese Renewable Energy Association]. December 2015. Available from http://www.apren.pt/ (in Portuguese).

Ecosativa. (2007). Parque Eólico da Serra de Bornes – Monitorização de Avifauna: 1º Relatório (Fase anterior à Construção) [Bornes Wind Farm—Bird Monitoring: Report 1 (Pre-Construction phase)]. Relatório para PESP, S.A. Portimão, Portugal (in Portuguese).

Eichhorn, M., & Drechsler, M. (2010). Spatial trade-offs between wind power production and bird collision avoidance in agricultural landscapes. *Ecology and Society, 15*(2), 10 [online] URL: http://www.ecologyandsociety.org/vol15/iss2/art10/

European Commission. (2011). *Wind energy developments and natura 2000—Guidance document: EU guidance on wind energy development in accordance with the EU nature legislation* (116 pp). Luxemburg: Publications Office of the European Union.

Fijn, R. C., Krijgsveld, R. C., Poot, M. J. M., & Dirksen, S. (2015). Bird movements at rotor heights measured continuously with vertical radar at a Dutch offshore wind farm. *Ibis, 157*(3), 558–566.

Fischer, R. A., Guilfoyle, M. P., Valente, J., Gauthreaux, S. A., Belser, C. G., Blaricom, D. V., et al. (2012). *The Identification of Military Installations as Important Migratory Bird Stopover Sites and the Development of Bird Migration Forecast Models: A Radar Ornithology Approach* (151 pp). SERDP Project SI-1439. Environmental Laboratory ERDC/EL TR-12-22. Prepared for U.S. Army Corps of Engineers Washington, DC 20314-1000. USA.

Fraga, M., Romero-Pedreira, D., Souto, M., Castro, D. & Sahuquillo, E. (2008). Assessing the impact of wind farms on the plant diversity of blanket bogs in the Xistral Mountains (NW Spain). *Mires and Peat, 4*: Article 6. Available from http://www.mires-and-peat.net

Gese, E. M. (2001). *Monitoring of terrestrial carnivore populations*. USDA National Wildlife Research Center—Staff Publications. Paper 576. Available from http://digitalcommons.unl.edu/icwdm_usdanwrc/576

GPWind. (2013). *Good practice guide—A new resource for reconciling wind energy development with environmental and community interests*. Available from https://ec.europa.eu/energy/intelligent/projects/sites/iee-pro-jects/files/projects/documents/gpwind_good_practice_guide_gp_wind_en.pdf

GWEC. (2016). *Global Wind Report 2015 – Annual market update*. Global Wind Energy Council. Brussels, Belgium. Available from http://www.gwec.net/publications/global-wind-report-2/global-wind-report-2015-annual-market-update

Hardey, J., Crick, H., Wernham, C., Riley, H., Etheridge, B., & Thompson, D. (2006). *Raptors: A field guide to survey and monitoring*. Edinburgh: Scottish Natural Heritage.

Horn, J., Arnett, E. B., & Kunz, T. H. (2008). Behavioral responses of bats to operating wind turbines. *Journal of Wildlife Management, 72*(1), 123–132.

Hull, C. L., & Muir, S. (2010). Search areas for monitoring bird and bat carcasses at wind farms using a Monte-Carlo model. *Australasian Journal of Environmental Management, 17*(2), 77–87.

Huso, M. (2010). An estimator of wildlife fatality from observed carcasses. *Environmetrics, 10* (22), 318–329.

Huso, M., & Dalthorp, D. (2014). Accounting for unsearched areas in estimating wind turbine-caused fatality. *The Journal of Wildlife Management, 78*(2), 347–358.

Huso, M. M., Dalthorp, D. H., Dail, D. A., & Madsen, L. J. (2015). Estimating wind-turbine caused bird and bat fatality when zero carcasses are observed. *Ecological Applications, 25*(5), 1213–1225.

ICNB. (2009). Recomendações para Planos de Monitorização de Parques Eólicos – Quirópteros [Recommendations for Monitoring Plans in Wind Farms]. Instituto para Conservação da Natureza e Biodiversidade [Institute for Nature and Biodiversity Conservation]. Portugal. Available from http://www.icnf.pt/portal/naturaclas/patrinatur/resource/docs/Mam/morc/morc-recom-p-eolic/ (in Portuguese).

ICNB. (2010a). *Orientações relativas à natureza e aplicação de Medidas de Compensação no contexto da aplicação do Decreto-Lei N.º 140/99, de 24 de Abril, Republicado pelo Decreto-Lei N.º 49/2005, de 24 de fevereiro. [Guidelines to defining and implementing Compensation Measures in the contexto Decree Law No. 140/99 of April 24th, amended by Decree Law No. 49/2005 of February 24th].* Instituto para Conservação da Natureza e Biodiversidade [Institute for Nature and Biodiversity Conservation]. Portugal. Available from http://www.icnf.pt/portal/naturaclas/ordgest/aa/resource/doc/med-comp-dez2010 (in Portuguese).

ICNB. (2010b). *Orientações para Monitorização dos Efeitos de Intra-Estruturas sobre o Lobo. [Guidelines to Monitoring Infraestruture Effects on Wolves].* Instituto para Conservação da Natureza e Biodiversidade [Institute for Nature and Biodiversity Conservation]. Portugal. Available from http://www.icnf.pt/portal/naturaclas/ordgest/aa/resource/doc/orientac-monit-lobo-aia (in Portuguese).

ICNB. (2010c). *Avaliação do efeito dos parques eólicos sobre os morcegos em Portugal continental - análise dos dados disponíveis em Outubro de 2009 [Evaluation of the effects of the wind farms on bats in continental Portugal].* Instituto da Conservação da Natureza e da Biodiversidade [Institute for Nature and Biodiversity Conservation]. Relatório não publicado (in Portuguese).

Jenkins, A. R., van Rooyen, C. S., Smallie, J. J., Harrison, J. A., Diamond, M., Smit-Robinson, H. A., et al. (2015). *Best—Practice guidelines for assessing and monitoring the impact of wind-energy facilities on birds in Southern Africa* (3rd ed., 67 pp). South Africa: Birdlife South Africa; Endangered Wildlife Trust.

Kikuchi. (2008). Adverse impacts of wind power generation on collision behaviour of birds and anti-predator behaviour of squirrels. *Journal for Nature Conservation, 16*, 44–55.

Korner-Nievergelt, F., Brinkmann, R., Niermann, I., & Behr, O. (2013). Estimating bat and bird mortality occurring at wind energy turbines from covariates and carcass searches using mixture models. *PLoS ONE, 8*(7), e67997. doi:10.1371/journal.pone.0067997.

Korner-Nievergelt, F., Korner-Nievergelt, P., Behr, O., Niermann, I., Brinkmann, R., & Hellriegel, B. (2011). A new method to determine bird and bat fatality at wind energy turbines from carcass searches. *Wildlife Biology, 17*(4), 350–363.

Kunz, T., Arnett, E. B., Cooper, B. M., Erickson, W. P., Larkin, R. P., Mabee, T., et al. (2007). Assessing impacts of wind-energy development on nocturnally active birds and bats: A guidance document. *Journal of Wildlife Management, 71*(8), 2449–2486.

Kuvlesky, W. P., Brennan, L. A., Morrison, M. L., Boydston, K. K., Ballard, B. M., & Bryant, F. C. (2007). Wind energy development and wildlife conservation: Challenges and opportunities. *Journal of Wildlife Management, 71*(8), 2487–2498.

Marchante, H., Morais, M., Freitas, H., & Marchante, E. (2014). Guia prático para a identificação de Plantas Invasoras em Portugal [Practical Guide to the Identification of Alien Plants in Portugal]. Coimbra: Imprensa da Universidade de Coimbra. Portugal. Available from https://issuu.com/plantasinvasoras/docs/guia_plantas__invasoras_em_portugal (in Portuguese).

Marques, A. T., Batalha, H., Rodrigues, S., Costa, H., Ramos Pereira, M. J., Fonseca, C., et al. (2014). Understanding bird collisions at wind farms: An updated review on the causes and possible mitigation strategies. *Biological Conservation, 179*, 40–52.

Marques, A. T., Paula, J., Marques, J., Pereira, M. J. R., Ramalho, R., & Rodrigues, S. (2015). Assessing the problem. In M. Mascarenhas, J. Bernardino, A. Paula, H. Costa, C. Bastos, A. Cordeiro, et al. (Eds.), *Biodiversity and wind energy: A bird's and bat's perspective* (96 pp). Portugal: Bio3 and University of Aveiro.

Marshall, R., Arts, J., & Morrison-Saunders, A. (2005). International principles for best practice EIA follow-up. *Impact Assessment and Project Appraisal, 23*(3), 175–181.

Masden, E. A., Haydon, D. T., Fox, A. D. Furness, R. W., Bullman, R., & Desholm, M. (2009). 2 Barriers to movement: Impacts of wind farms on migrating birds. *ICES Journal of marine Science, 66*(4), 746–753.

Mathews, F., Swindells, M., Goodhead, R., August, T., Hardman, P., Linton, D. M., et al. (2013). Effectiveness of search dogs compared with human observers in locating bat carcasses at wind-turbine sites: A blinded randomized trial. *Wildlife Society Bulletin, 37*(1), 34–40.

NRC. (2007). *Environmental impacts of wind-energy projects.* National Research Council of the National Academies. The National Academic Press, Washington, D.C., USA.

Orloff, S., & Flannery, A. (1992). *Wind turbine effects on avian activity, habitat use and mortality in Altamont Pass and Solano county wind resource areas: 1989–1991.* Final Report. California Energy Commission, USA.

Parsons, S., & Szewczak, J. M. (2009). Detecting, recording, and analysing the vocalizations of bats. In T. H. Kunz & S. Parsons (Eds.), *Ecological and behavioral methods for the study of bats* (2nd ed., pp. 91–111). Baltimore: Johns Hopkins Press.

Passos, I., Silva, M., Mesquita, S., Marques, A., Bernardino, J., Costa, H., et al. (2013, February). Aliens in wind farms—Preventing and monitoring impacts on vegetation. In *Communication presented at Conference on Wind Energy and Wildlife Impacts.* Stockholm, Sweden. Available from http://www.naturvardsverket.se/CWE2013/Emerging-issues/

Paula, A., Santos, J., Cordeiro, A., Costa, H., Mascarenhas, M., & Reis, C. (2011b, May). *Habitat management for prey recovery: An off-site mitigation tool for wind farms impacts on top avian predators.* Communication Presented at the Conference on Wind energy and Wildlife impacts, Trondheim, Norway. Available from http://cww2011.nina.no/

Paula, A., Santos, J., Ferreira, C., Bernardino, J., & Costa, H. (2015a). Compensation scenarios to deal with wind farm's impacts on birds: The challenges of moving from theory to practice. In *Conference of Wind energy and Wildlife Impacts—Book of Abstracts* (p. 51). Berlin, Germany. Available from https://www.cww2015.tuberlin.de/fileadmin/fg123_windsynopse/cww2015/downloads/CWW2015_Book_of_Abstracts.pdf

Paula, J., Bispo, R., Leite, A., Pereira, P., Costa, H., Fonseca, C., et al. (2015b). Camera-trapping as a methodology to assess the persistence of wildlife carcasses resulting from collisions with human-made structures. *Wildlife Research, 41*(8), 717–725.

Paula, J., Leal, M. C., Silva, M. J., Mascarenhas, R., Costa, H., & Mascarenhas, M. (2011a). Dogs as a tool to improve bird-strike mortality estimates at wind farms. *Journal for Nature Conservation, 19,* 202–208.

Pearce-Higgins, J. W., Stephen, L., Langston, R. W., Bainbridge, I. P., & Bullman, R. (2009). The distribution of breeding birds around upland wind farms. *Journal of Applied Ecology, 46,* 1323–1331.

Peste, F., Paula, A., Silva, L. P., Bernardino, J., Pereira, P., & Mascarenhas, M. (2015). How to mitigate impacts of wind farms on bats? A review of potential conservation measures in the European context. *Environmental Impact Assessment Review, 51,* 10–22.

Phillips, S., & Dudík, M. (2008). Modeling of species distributions with Maxent: New extensions and a comprehensive evaluation. *Ecography, 31*(2), 161–175.

Plecotus. (2008). *Parque Eólico da Serra de Bornes – Monitorização de Quirópteros: 1º Relatório – Ano 2005 (Fase anterior à Construção) [Bornes Wind Farm—Bat Monitoring: Report 1—Year 2005 (Pre-construction phase)].* Relatório para PESP, S.A. Pombal, Portugal (in Portuguese).

Richardson, D. M., Pysek, P., Rejmánek, M., Barbour, M. G., Panetta, F. D., & West, C. (2000). Naturalization and invasion of alien plants: Concepts and definitions. *Diversity and Distributions, 6,* 93–107.

Rodrigues, L., Bach, M. J., Dubourg-Savage, B., Karapandza, D., Kovac, T., Kervyn, J., et al. (2015). *Guidelines for consideration of bats in wind farm projects—Revision 2014* (133 pp). EUROBATS Publication Series No. 6 (English Version). UNEP/EUROBATS Secretariat, Bonn, Germany.

Rosa, I. M. D., Marques, A. T., Palminha, G., Costa, H., Mascarenhas, M., Fonseca, C., et al. (2016). Classification success of six machine learning algorithms in radar ornithology. *Ibis, 158,* 28–42.

Santos, M., Bastos, R., Travassos, P., Bessa, R., Repas, M., & Cabral, J. A. (2010). Predicting the trends of vertebrate species richness as a response to wind farms installation in mountain ecosystems of northwest Portugal. *Ecological Indicators, 10,* 192–2015.

Santos, H., Rodrigues, L., Jones, G., & Rebelo, H. (2013). Using species distribution modelling to predict bat fatality risk at wind farms. *Biological Conservation, 157,* 178–186.

Sea Green Wind Energy. (2011). *Seagreen Phase 1: Onshore Grid Connection Works Scoping Report.* Reporting to the Project Round 3 Zone 2—Firth of Forth Offshore Wind Farm Development.

Schuster, E., Bulling, L., & Köppel, J. (2015). Consolidating the state of knowledge: A synoptical review of wind energy's wildlife effects. *Environmental Management, 56*(2), 300–331.

Silva, M. & Passos, I. (2017). Vegetation. In M. Perrow (Ed.), *Wildlife and wind farms—Conflicts and solutions,* Vol. 1: Onshore—Potential effects (298 pp). United Kingdom: Pelagic Publishing.

Silva, M., Passos, I., Bernardino, J., Costa, H., & Mascarenhas, M. (2011, May). *Impacts on vegetation: A minor detail from a bigger problem*? Poster Presented at Conference on Wind Energy and Wildlife Impacts, Trondheim, Norway. Available from http://cww2011.nina.no/Posters.aspx

Singh, K., Baker, E. D., & Lackner, M. A. (2015). Curtailing wind turbine operations to reduce avian mortality. *Renewable Energy, 78,* 351–356.

Smallwood, K. S. (2007). Estimating wind turbine-caused bird mortality. *The Journal of Wildlife Management, 71*(8), 2781–2791.

Smith, E. P. (2002). *BACI design*. Encyclopedia of environmetrics (Vol. 1, pp. 141–148). Chichester, United Kingdom: Wiley Ltd.

SNH. (2009a). *Monitoring the impact of onshore wind farms on birds*. Guidance Note. Scottish Natural Heritage. Scotland, UK. Available from http://www.snh.gov.uk/

SNH. (2009b). *Guidance on methods for monitoring bird populations at onshore wind farms.* Guidance Note. Scottish Natural Heritage. Scotland, UK. Available from http://www.snh.gov.uk/

SNH. (2012). *Assessing the cumulative impact of onshore wind energy developments.* Guidance. Scottish Natural Heritage. Scotland, UK. Available from http://www.snh.gov.uk

SNH. (2014). *Recommended bird survey methods to inform impact assessment of onshore wind farms*. Guidance. Scottish Natural Heritage. Scotland, UK. Available from http://www.snh.gov.uk/

Sowler, S., Stoffberg, S., MacEwan, K., Aronson, J., Ramalho, R., Potgieter, K. & Lötter, C. (2016). *South African Good Practice Guidelines for Surveying Bats at Wind Energy Facility Developments - Pre-Construction*. (4th ed.). South African Bat Assessment.

Strickland, M. D., Arnett, E. B., Erickson, W. P., Johnson, D. H., Johnson, G. D., Morrison, M. L., et al. (2011). *Comprehensive guide to studying wind energy/wildlife interactions*. Washington, D.C., USA.

STRIX. (2015). *Radar assisted shutdown on demand for wind farms in Egypt*. Lisbon: Meeting Market Egypt.

Tomé, R., Leitão, A. H., Canário F., Pires, N. (2012). *Environmental management of wind farms—A successful onshore case study*. FAME international Workshop—Effects of Marine Renewables and other marine uses on Biodiversity—Atlantic Area.

Vilà, M., Espinar, J., Hejda, M., Hulme, P., Jarosık, V., Maron, J., et al. (2011). Ecological impacts of invasive alien plants: A meta-analysis of their effects on species, communities and ecosystems. *Ecology Letters, 14,* 702–708.

Williams, B. (2011). Adaptive management of natural resources—Framework and issues. *Journal of Environmental Management, 92,* 1346–1353.

WWEA. (2016). *WWEA quarterly bulletin—Wind energy around the world*. World Wind Energy Association. Issue 1 March 2016.

Chapter 4
An Overview on Methods to Assess Bird and Bat Collision Risk in Wind Farms

Sandra Rodrigues, Luís Rosa and Miguel Mascarenhas

4.1 Introduction

It is well known that wind farms present several impacts on wildlife, especially in flying vertebrates such as birds and bats (e.g.: Drewitt and Langston 2006; Percival 2005; Barrios and Rodríguez 2004; Tellería 2008; SPEA 2005; Johnson et al. 2002; Krijgsveld et al. 2009). However, the scientific community, as well as environmental practitioners, have focused their attention on fatality impacts. Understanding why different biological groups and species suffer higher levels of fatalities than others is essential to improving impact prediction and implementing more efficient measures to reduce fatalities (Chamberlain et al. 2006; BTO 2005; Drewitt and Langston 2006; Marques et al. 2014). The probability of collision (hereafter designated collision risk) relies on many different factors such as species' intrinsic characteristics, environmental conditions and even the features of wind farms (Drewitt and Langston 2006; Marques et al. 2014). Therefore, an increasingly variety of methods have been developed to assess this risk of collision in wind farms.

Collision risk models (known as CRMs) aim to provide a prediction of the collision risk that might be useful as a support for the decision-making process, prior to the construction phase, regarding permitting, mitigation and micrositing

S. Rodrigues (✉) · L. Rosa · M. Mascarenhas
Bioinsight, Odivelas, Portugal
e-mail: sandra.r@bioinsight.pt

L. Rosa
e-mail: luis.r@bioinsight.pt

M. Mascarenhas
e-mail: miguel.m@bioinsight.pt

S. Rodrigues
Department of Statistics and Operational Research, Faculty of Sciences, University of Lisbon, Lisbon, Portugal

M. Mascarenhas et al. (eds.), *Biodiversity and Wind Farms in Portugal*,
https://doi.org/10.1007/978-3-319-60351-3_4

(Normandeau 2012). Most of these tools were developed for the post-construction phase, when mortality estimation is not possible, as in the case of offshore developments. CRM are heuristic tools that consist of an explicit mathematical or algorithmic structure that generates output data in an automated fashion, that predict system outcomes, based on a given starting condition, defined by model input data (Willmott et al. 2012). They usually include a calculation of the probability of a collision occurring and an estimation of the number of birds within a risk window in order to estimate the likely number of collision events (Masden and Cook 2016) and can be separated into two basic types depending on the way in which they are constructed: (a) empirical models, that are built starting with collected data and observed patterns (Willmott et al. 2012); and, (b) mechanistic models that are built starting with mathematical representations of the dynamics of the studied system (Tham 2000 in Willmott et al. 2012).

Other risk assessment-related tools that are important in the decision-making process of wind farm projects involve vulnerability indexes. These indexes rank the species that are more sensitive to collision and fatality impact and usually include factors such as conservation status and population trend, as well as flight characteristics. These are important tools in the decision-making process, from a conservation point of view, since understanding which species will have higher risk of collision is not the only important factor. Certain species may have a very low risk of collision, which translates in low fatalities at wind farms; however, even low mortality rates can put extremely vulnerable populations at conservation risk and affect the capability of population restoration and survival (Drewitt and Langston 2006; Carrete et al. 2009). The possibility to foresee which these species are, by means of a vulnerability index supported by statistical models that combine the species' vulnerability to collision and conservation vulnerability, is essential for species conservation. It will provide crucial indications on which species should be considered as targets during wind farm impact assessments and monitoring studies.

Despite the approach, it must be taken in consideration that models and indexes are approximations to the reality, and it is important to validate them and define how robust and sensible they are to the violations of their assumptions. Nevertheless, the literature shows that CRMs are rarely validated (Masden and Cook 2016; Willmott et al. 2012).

In this chapter, we intend to provide a literature review on a range of CRMs as well as other methodologies for collision risk assessment used internationally and in Portugal. We then analyse each methodology and discuss their potential to be used in the Portuguese environmental consultancy context, taking in consideration its limitations, advantages, result validation, and difficulties to implement (such as complexity of the input data and sampling effort requirements). Methodologies that have the best results and show some room for improvement are highlighted. We suggest where future efforts should be made to advance collision risk assessment and modelling into the decision-making process of wind farm projects, mainly in terms of the Portuguese reality. We then present a case study of a widely used tool in Portugal for avian risk assessment and its application in a Portuguese wind farm.

4.2 Collision Risk Models

Masden and Cook (2016) made an extensive review of the literature on avian collision models in peer-reviewed scientific literature as well as grey literature. Following their work and for the present review, 10 CRM were considered and are listed below. Other models are also available; however, they lack accessible documentation, with some models being proprietary and under commercial confidentially, and therefore unreachable for consultation. The models identified in the literature and taken under consideration for this review are:

- CRM by Tucker (Tucker 1996a, b)
- CRM by Band et al. (2007)—first developed by the Scottish Natural Heritage (SNH 2000)
- CRM by McAdam (2005)
- CRM by Podolsky (2008)
- Hamer model (Holmstrom et al. 2011)
- Biosis model (Smales et al. 2013)
- CRM by (Eichhorn et al. 2012)
- CRM by Desholm (2006)
- CRM by Bolker et al. (2014)
- U.S. Fish and Wildlife Service model (USFWS 2013)

All the considered CRMs include an estimate of the probability of collision with a wind turbine of a single bird while flying through the rotor swept area. However, they differ in terms of how the estimation is made and in the degree of complexity of the calculations, and rely upon previous knowledge on bird and wind turbine characteristics. In general, it is possible to say that, despite the crescent number of models developed in the last decade, CRM models comprise at least one or two main aspects: (i) the number of birds exposed to turbines and, therefore, at risk of collision; and (ii) the probability of an individual bird colliding. Most of the differences between CRMs relate to the former rather than the latter (Masden and Cook 2016).

In the next paragraphs, a short description of the models are presented and discussed.

4.2.1 Tucker Model (Tucker 1996a)

Tucker (1996a) was the first to publish an analysis for bird–rotor collisions in wind farms. He developed a mechanistic model that calculates a probability of collision when birds fly through the rotor swept zone, varying according to bird and turbine motions and dimensions. It does not estimate a probable number of collisions, though a measure of bird density through the turbine is not included. The probability of collision is calculated as a ratio of the time taken for a bird to move through the rotor swept zone compared to the time taken for the turbine blades to complete a single revolution. It is considered that the bird always moves perpendicular or parallel to the turbine rotor but flight can be parallel or oblique to the wind direction and the model can accommodate upwind or downwind flight. Collision with the static elements of the turbine and avoidance behaviour of the bird, unlike some other models later developed, are not included in this model.

4.2.2 Band Model (Band et al. 2007)

The Band model (Band et al. 2007), first developed by the Scottish National Heritage (SNH 2000), is the most used model in the UK and is standard in many countries, being the most well-known and openly documented wind–wildlife collision risk model (Willmott et al. 2012). This is an empirical model that combines field study and statistical modelling, which can be used to predict collision mortality risk of birds at wind facilities (Band et al. 2007). Like Tucker (1996a), this model is also based on the probability of a turbine blade occupying the same space as a bird during the time it takes the bird to pass through the rotor swept volume of the turbine. The probability of collision relies on information about the bird (wing span, body length, flight speed, flight height, nocturnal flight activity) as well as the turbine (blade width, blade length, blade pitch, rotor speed, hub height, operational time) (Band 2012b). The collision prediction of the original, or basic, model has two key components for estimating the number of collisions per annum: (i) the number of birds flying through the rotor; and (ii) the probability of collision from a single flight through the rotor. The most recent version also includes a measure of avoidance behaviour, allowing consideration for a proportion of birds to avoid collision (Band et al. 2007). The extended model is built upon the basic model, calculating the probability of a bird flying at a particular height within the turbine rotor swept zone using flight height curves (Johnston et al. 2014), while the basic model assumes a uniform distribution of birds across the rotor swept zone of the turbine. The individual probabilities are then integrated to gain the collision integral. The Band model has been further developed for application in offshore environments (Band 2012a, b).

4.2.3 Other Tucker/Band Based Models

Several other authors built upon the two precursor models mentioned above (Tucker and Band models), adopting different approaches and adding different parameters, reflecting the complexity and various numbers of factors that are known to influence the probability of collision (see Drewitt and Langston 2006; Marques et al. 2014).

The McAdam model (2005) is based on the original Band model (SNH 2000), but takes the variation of flight height and wind effects into account. It includes oblique angles of the approach of the bird, but not the variation in bird orientation that is relative to the turbine (Masden and Cook 2016). The Podolsky model (2008) follows similar methods as Band et al. (2007) and Tucker (1996a) and also considers different angle flights approaches, adding a proportion of birds that avoid collision (avoidance behaviour) to the model; Holmstrom et al. (2011), also based on Tucker (1996a, b) and also considering the different oblique angles of approach, however, ignored avoidance behaviour.

In the biosis model, Smales et al. (2013) considered the static components of the generator as others had done, but considered the possibility that birds could approach turbines from any direction. It already added turbine layout variable to the model, and estimated the sum of the average number of turbine encountered per flight within a scattered wind turbine array, rather than for all turbines in an array, calculating it using a topological non-affine mapping technique (Masden and Cook 2016).

Eichhorn et al. (2012) used the probability of the collision component from the Band model in their individual-based model of *Milvus milvus*' foraging. This model approaches collision risk as a function of distance to turbines, nest and other parameters. The probability of collision, using the second component of the Band model (the probability of collision from a single flight through the rotor), is calculated when a bird occurs in the area near a wind turbine.

Other models developed by Desholm (2006) and Bolker et al. (2014) assume the probability of collision is fixed and known; therefore, they use a constant probability of collision instead of calculating a variable probability. The use of a single constant probability removes the theoretical calculation of the probability of collision as presented by Tucker (1996a) for example, and the associated uncertainty of factors such as avoidance (Masden and Cook 2016). Desholm (2006) includes an array of many 'new' variables, such as: information on bird migration volume, proportion of birds entering the wind farm, proportion of birds within the horizontal range of the rotor- lades, proportion of birds in the vertical range of the rotor blades, proportion of birds trying to cross the rotor swept zone without showing avoidance, the mean number of turbine rows that were already crossed by the bird, and the probability of passing safely through the rotor just by chance, which was assumed to be fixed and taken from Tucker (1996a) rather than calculated directly. Additionally, it also includes information on the layout of the wind farm as a factor for risk of collision: Bolker et al. (2014) analysed the geometry of the wind farm, and determined the average number of turbines encountered if birds moved through the wind farm, assuming said birds flew in straight lines.

4.2.4 The Bayesian Approach Model

As a Bayesian alternative to the previous models, the model developed by the U.S. Fish and Wildlife Service (USFWS 2013) does not calculate the probability of collision mechanistically, while allowing for the estimate of probability of collision to be amended and updated as information becomes available (Masden and Cook

2016), following the principles of a Bayesian statistical approach. The model was developed for eagle species and the total annual estimated eagle fatalities is the product of the rate of eagle exposure to turbines (eagles present in the area of interest per hour per square km), the probability that eagle exposure will result in a collision with a turbine (based on SNH (2009) but any suitable data and studies could theoretically be used), and an expansion factor that scales the fatality rate to a predicted number of annual fatalities, based on information on the number of turbines and the hazardous area surrounding a turbine (Masden and Cook 2016). The model, while not including detail on the mechanisms of collision, allows for adaptive management and updating of information as it becomes available, such as actual collision events, given the Bayesian nature of the model (Masden and Cook 2016).

4.3 Vulnerability Indexes and Other Tools for Collision Risk Assessment

Other tools have been identified in the literature and are in use for collision risk assessment, as an alternate approach to the CRM. They usually come in form of an index, but others, such as algorithms-based models and GIS-based tools are presented. The tools identified in the literature and taken under consideration for this review are:

- Collision hazard index (CHI) (Bio3 2008)
- Vulnerability index by Garthe and Hüppop (2004)
- Vulnerability index by Furness et al. (2013)
- Spatial vulnerability index (Noguera et al. 2010)
- Avian species prioritisation species (Beston et al. 2016)
- Fatality risk index for Iberian avian species (Soares 2014)
- INTACT (INnovative Tools to reduce Avian Collisions with wind Turbines) micrositing GIS tool (Hanssen and May 2015)
- ProBat tool (WindBat 2016)—the tool to calculate bat friendly operating periods.

In the next paragraphs, a short description of the tools are presented and discussed.

4.3.1 The Collision Hazard Index (CHI) (Bio3 2008)

The CHI is a method used in Portugal and developed for raptors and other soaring birds, as a response to the wind energy expansion in the country at the date and the increasing necessity to use tools that interpret bird's behaviour as a collision risk. It was then enhanced in the following years (refer to examples: Bio3 2008, 2009a,

2014a). With slight variations between each monitoring project, this methodology is usually based on field data collected throughout the year from several vantage points across the wind farm. Avian routes are recorded according to the several categories of flight behaviour and height, representing different levels of risk. The field data collected is analysed with a GIS tool using a square grid of the area to display the information of collision risk (see Sect. 4.4.1 for more information). This tool is particularly useful in pre-construction phase, allowing the identification of areas most used by sensitive species and where the flights represent most risk of impact, and aids in decision-making process and in the definition of no-go areas. This tool is also used in post-construction phase, to help establish the areas and species that deserve some attention during the monitoring plans.

4.3.2 Vulnerability Indexes

Authors such as Garthe and Hüppop (2004), Noguera et al. (2010), Furness et al. (2013) and Beston et al. (2016) developed vulnerability indexes. This type of tool can be implemented for different groups of birds in order to reach a better understanding of the real impact in each species populations, based on each species' vulnerability to collision and conservation sensitivity. These tools focus on species characteristics in order to determine which morphological and ethological factors better explain bird and bat species proneness to collide with wind farm turbines, framing it with their conservation status or population trend.

Garthe and Hüppop (2004) developed a wind farm sensitivity index for seabirds using nine factors scored on a point scale, based on fields assessments and existing knowledge or author's subjective considerations. Furness et al. (2013) followed a similar approach incorporating new data and modifying the allocated scores, based on updated research. Noguera et al. (2010) adapted the index developed by Garthe and Hüppop (2004) to raptors species in onshore areas.

More recently, Beston et al. (2016) developed a prioritisation system to identify the avian species most likely to experience population declines from wind facilities based on their current conservation status and their expected risk from turbines. They developed an approach that determines collision risk associated with each turbine, which incorporates (i) data on collision fatalities at wind facilities; (ii) population size; (iii) life history; (iv) species' distributions relative to turbine locations; (v) number of suitable habitat types; and (vi) each species' conservation status. This prioritisation system highlights both species that could potentially experience population decline caused by wind energy and species at low risk of population decline (Beston et al. 2016).

A similar approach to Noguera et al. (2010), was developed for the Portuguese context (Soares 2014; Bispo et al. 2015). This is a model-based index that indicates each species' proneness to collision and weighs its sensitivity from the conservation point of view, integrating the species conservation status, population trend and breeding capacity. This was constructed using observed mortality data collected under an array of Portuguese wind farms monitoring programmes. Although not in use in Portugal at the moment, the implementation of vulnerability indexes adapted to the present communities is very relevant since the majority of wind farms are located in areas of ecological concern (SPEA 2005).

4.3.3 INTACT Micrositing GIS-Tool (Hanssen and May 2015)

A GIS-based micrositing tool that addresses the collision risk from a new perspective is currently being developed as part of the INTACT project (INnovative Tools to reduce Avian Collisions with wind Turbines). Birds have different morphology and behavioural characteristics that influence species-specific flight strategies, as well as species-specific capabilities to exploit orographic and thermal updrafts as they traverse the landscape. This tool tackles the problem of the influence of landscape orientation and thermal and orographic updrafts in soaring bird flights in mind. Soaring birds are attracted by topography, since it can provide local updrafting winds to subsidise soaring and gliding flight activity. Soaring birds do not maintain high altitudes during migration. Instead, the birds conserve energy by gaining lift from updrafts and gliding long distances, slowly losing altitude, to the next updrafts (Newton 2007; Kerlinger 1989; Katzner et al. 2012). During migration, raptors prefer to follow topographical 'leading lines' such as ridges, mountain ranges and main river systems. Such orientation features depends on the geomorphometric complexity of the entire landscape at different scales in turn affecting how birds prefer to migrate across the landscape (Hanssen and May 2015). The tool was developed with the objective of identifying high-risk turbine locations for birds in relation to thermals, local topography and wind conditions, rather than focusing in turbine characteristics. While the previous models address the issue from the moment birds are in the turbines radius of action, the Hanssen and May (2015) methodology in the INTACT tool analyses the factors that bring birds closer to turbine areas, leading them to a situation where they are at risk. The output of this model is in the form of an intuitive map of specific avian interest, and highlighted no-go areas.

4.3.4 ProBat Tool (WindBat 2016)

The ProBat tool, is the only bat-related tool found in literature that can somehow be related to bat risk assessment; however, it is not currently used as a risk assessment tool. It was developed in Germany by the research project RENEBAT, now in its third version (RENEBAT III) (WindBat 2016). The project's research team developed an algorithm-based method and a toolbox for predicting bat activity and the risk of collision of bats detected in the rotor area of wind turbines (FAU and LSE n.d.). It is used in Germany, by many parties involved in wind farms, as a standardised protocol solution for real time bat mitigation measure since it allows calculation of specific starting wind speeds for bat-friendly operation of the wind farm. The developed toolbox is intended to facilitate the implementation of the algorithm for the calculation of bat-friendly operating periods and is intended to be a pre-requisite for a fast, inexpensive and environmentally compatible expansion of the wind energy. While its use is mainly at the operational phase as a mitigation measure, we have considered that there might be room for adaptation to the pre-construction phase for estimation of bat risk periods, since the tool allows prediction of periods of shutdown turbines due bat high risk levels if bat activity data (in this case pre-construction monitoring bat activity data) is introduced. This data was originally intended as a means to estimate the loss of economic income as a result of the operating algorithms (WindBat 2016).

4.4 Methodologies Comparison: Advantages and Limitations

4.4.1 Collision Risk Models

One of the biggest limitations of CRMs, much like other methodologies used to predict collisions, relies on the assumptions they make about bird behaviour. Most of the models assume a linear relationship between bird abundance and collision risk which may not always be verified (de Lucas et al. 2008). Some authors (Barrios and Rodríguez 2004; Schaub 2012; De Lucas et al. 2012) have shown that interactions between topography, species-specific behaviour, turbine layout, and even wind turbine features are likely to exist. The most common limitations identified in the literature are:

- The collision risk being defined as constant in the whole rotor swept zone, when it is actually known to be variable (Cook et al. 2014) and under the influence of other variables as behaviour or wind;
- Including avoidance behaviour in the form of an avoidance rate that assumes that a certain proportion of those birds on a collision path will adopt this behaviour. Most models assume that avoidance behaviour is constant across all individuals within a species, which is unlikely;
- Lack of solid data on avoidance rates. Estimating variability between species is difficult, least of all within a species (Cook et al. 2014). Nevertheless, UHI (2015) has considered that avoidance rate is not one of the most important factors. Instead, velocity and flight height are considered to be more important factors by UHI (2015);
- Lack of validation. This is pointed out as one of the main fragilities of CRM models (Willmott et al. 2012; UHI 2015; Masden and Cook 2016);

- UHI (2015) points out that there is a need for a greater acceptance of the uncertainty of the models. Models are not supposed to be definitive and black and white, therefore, context on their uncertainty should be provided by whom uses them. It is also pointed out that the modelling process itself should be more reproducible, and uncertainty should be factored into estimates;
- They are determinist models, excluding variability or uncertainty from calculations. The incorporation of uncertainty in future models would reduce the possibility that a collision estimate was driven by the choice of a single input parameter value (Masden and Cook 2016).

On the other hand, simpler models but within a Bayesian framework (USFWS 2013) are an alternative approach to the increasingly complex mechanistic models for collision estimation. Such models can accommodate increasing complexity, but have a drawback: in the absence of knowledge, these models remain simple, using empirical data on actual collision events to determine parameter values (Masden and Cook 2016). Therefore, these models rely on the ability to collect data on actual collisions in order to be validated. However, they have the advantage of being able to be progressively better if they are constantly feed with more and more data.

4.4.2 Vulnerability Indexes and Other Tools for Collision Risk Assessment

The CHI (Bio3 2008) is a straightforward and intuitive way of addressing the collision risk, but it is too site-specific as it does not allow for conclusions to be reached about the variables responsible for collisions. Additionally, it only considers that the same flight types and heights represent the same risk value for different species or different species groups and does not consider differences between species. It might be considered to be too simplistic, but, while having room for improvement, it has proved to be a useful tool to address collision risk, and the post-construction mortality monitoring and observed fatalities in wind farms in Portugal confirms, to some degree, the results obtained with this methodology (see Sect. 4.4.1).

The use of vulnerability indexes can give the species-specific input that most collision indexes lack (e.g., CHI by Bio3 2008; Spatial vulnerability index by Noguera et al. 2010). The avian species prioritisation species (Beston et al. 2016) and the Fatality Risk Index for Iberian avian species (Soares 2014) have the advantage to identify the species characteristics that might be driving the collisions. Additionally, it provides a rank of species that are most likely to be affected and that have a bigger risk of collision, for the country the database is from. In addition, it is updatable as new mortality data is obtained; for example, the use of the Iberian avian vulnerability index (Soares 2014) in integration with the CHI could greatly enhance wind farm collision risk assessments. The vulnerability index, by prioritising species, could help to direct efforts to those more sensitive and so, more relevant to decide upon.

The INTACT Micrositing GIS-tool (Hanssen and May 2015) can represent a realistic measure of potential harmful turbine locations in complex terrain. The down side is that it is still a tool under development and which involves complex modulation. Updrafts, especially thermal updrafts, are very difficult to modulate and data is not easily available. It also focuses on updraft flights and ignores other variables that might be important. Another limitation is that the tool focuses on species that use updrafts in their flights, mainly birds of prey. The advantage is that the output of the model is a generalist map with highlighted no-go areas, which are excellent for tools for site selection and turbine micrositing.

In Table 4.1, we summarise a comparison of different tools' characteristics, advantages, limitations, inputs and outputs, as well as their field sampling requirements and decision-making utility, with the Portuguese reality in mind. Since offshore wind energy is yet to be developed in Portugal, we selected for comparison, only the models more prone to be applicable in an onshore environment. It is not a surprise that the INTACT micrositing GIS-tools (Hanssen and May 2015) and the CHI (Bio3 2008) are highlighted as the most useful tools for the decision-making process, mostly due to their outputs as intuitive maps, which make them very useful for site selection and micrositing of turbines, as well as in terms of their low requirements for field work. On the other hand, the models by Band et al. (2007) and Podolsky (2008) would potentially be very useful for the decision making process, but their intensive field sampling requirements are a drawback for use in the Portuguese setting (see Sect. 4.4 for more details), diminishing their usefulness.

4.5 The Portuguese Context

Despite the current boom and advances in collision risk models, mainly in the last decade, CRM tools are still to be implemented in Portugal. This might be explained by the legislation context and requirements for the project phase in Portugal. While some countries' policies tend towards an investment in extensive surveys before wind farm project approval, which allows a baseline scenario to be properly defined, in Portugal, the time window for that survey is narrower. Therefore, the operation phase gains most of the attention and turns out to be the main focus of analysis, due to extensive empirical monitoring. On the other hand, Portuguese wind farms projects are often small compared with other countries (average of 10 turbines per wind farm; Marques et al., Chap. 1), therefore impact assessments costs cannot get too high due to exaggerated levels of field effort/expertise, at the risk of projects in the wind energy sector becoming economically unviable. Additionally, despite the merits of the use of CRMs, these models carry some limitations and specificities that discourage an extended use in the Portuguese environmental impact assessment (EIA) process, as discussed in Sects. 4.2 and 4.2.1, such as the complexity of the data required to feed the models, leading to a requirement for large time investments, unadjusted for the required

Table 4.1 Avian collision risk assessment methodology comparison

	Methodology		Advantages	Limitations	Inputs	Outputs	Validation	Decision making utility	Field sampling requirements
In use internationally	Main onshore Collision Risk Models	Tucker model (Tucker 1996a, b)	Numerous bird and turbine parameters considered	Complex and mechanistic; outputs not useful for decision making; Week relation between predictions and mortality observed	Rotor and bird characteristics	Probability of collision	No	+	+++
		Band model (Band et al.2007)	Open document/easy access; Easy calculations; Many environmental and structural parameters	Intensive field sampling; Week relation between predictions and mortality observed	Bird field data and flight characteristics; Wind turbine dimensions	Number of birds colliding	No	++	+++
		Podolsky model (Podolsky 2008)	Simulates influences of parameters on collisions; Numerous bird and turbine parameters considered	Intensive field sampling; Proprietary; Model does not identify high risk areas; Week relation between predictions and mortality observed	Rotor and bird characteristics; Angle of approach and avoidance behaviour	Probability of collision	No	++	+++
		USFWS Eagle CM (USFWS 2013)	Bayesian framework: allows for the estimate of probability of collision to be amended and updated when information becomes available	Collision avoidance measures are difficult to estimate; Does not identify high-risk turbines	Rate of eagle exposure to turbines and probability that it will result in a collision; Expansion factor that scales the fatality rate to a predicted number of annual fatalities	Number of birds colliding	No	+	++

(continued)

Table 4.1 (continued)

	Methodology		Advantages	Limitations	Inputs	Outputs	Validation	Decision making utility	Field sampling requirements
In use or under consideration both internationally or in Portugal	Indexes and other tools for collision risk assessment	INTACT Micrositing GIS-tool (Hanssen and May 2015)	Intuitive maps of specific avian interest with highlighted no-go areas; Micrositing prone	Difficult to obtain thermal updrafts data; Raptor orientated; Generalistic output	Orographic and thermal updrafts models; Topografical and hydrological orientation; Relative value for each factor	Normalised risk map	No	+++	0
		Vulnerability Indexes (Garthe and Hüppop 2004; Noguera et al. 2010; Furness et al. 2013; Soares 2014; Beston et al. 2016)	Ranks the most vulnerable species; Complement of other methodologies in decision-making process	Usually mortality data dependent. Works mainly as a complement for other methodologies	Mortality data; Species characteristics and sensibility (conservation status; population trend; reproductive success)	Species risk rank	No	NA	Database dependent
In use in Portugal		Collision hazard index (Bio3 2008)	Easy to implement; Broad analysis of the study area; Micrositing prone	Simplistic; only behaviour oriented	Flight types, heights and routes	Use intensity map	No*	+++	++

Relevant methodologies are analysed with regards to their **Advantages** and **Limitations**, **Input** and **Output** data, if outputs were validated (**Validation**), usability for the decision-making process (**Decision making utility**) and required effort during field sampling (**Field sampling requirements**). For the analysis of the usability of each methodology for the decision-making process (Decision making utility), four categories were used: +++ very useful, ++ averagely useful, + low usefulness, *NA* Not applied. For the analysis of the effort required during field sampling (Field sampling requirements), five categories were used: 0 no field sample required, +++ requires great effort; ++ requires medium effort; + requires low effort; Database dependent—this means that although the methodology does not require field data, it is dependent on data from previous work. (*—indicates that although this tool is not statistically validated, results indicate a relation between areas where the tool indicated higher risk levels and fatalities events; see Cordeiro et al. 2013)

licencing in the present Portuguese context. On the other way, lack of result validation, might also discourage the use of CRM and hinder its widespread use in the country. Therefore, at the present time in Portugal, methodologies that are somewhat easier to implement, such as indexes, have been shown to be sufficient to comply with the requirements and objectives of wind farms' ecological risk evaluation.

In Portugal, wind farms have been exclusively onshore, with only two offshore prototypes currently in implementation until now. It is, however, expected that as the offshore industry slowly develops in the country, that Portuguese investigators and environmental practitioners will be motivated to develop and adapt tools to the challenges of offshore wind farm and risk assessments. Therefore, some of these internationally used tools might gradually gain importance within Portugal.

In the next section, a case study is presented demonstrating the use of a simple index to estimate collision risk—the CHI, developed in Portugal and in use since 2007 (see Bio3 2008). Its application at a Portuguese wind farm to estimate areas of higher risk of collision for the common kestrel (*Falco tinnunculus*) population inhabiting the area is discussed.

4.5.1 Case Study|Estimating Avian Collision Risk in Portugal: The CHI—A Practical Methodology

The CHI is an example of a simple approach that has been applied in Portugal since 2007 (refer to examples: Bio3 2008, 2009a, 2014a), mainly to understand raptors collision risk behaviour at wind farms (Marques et al. 2010). This analysis is both a practical and effective approach to avian impact assessment at wind farms and can be applied during the pre-construction and operational phase.

The methodological protocol behind CHI is based on a spatial analysis of avian routes, conducted on a 3D basis, reflecting the vertical and horizontal use of air space at an existing/proposed wind farm area.

4.5.1.1 The Collision Hazard Index

Figure 4.1 summarises the necessary steps for undertaking a CHI analysis, starting in a desktop analysis, followed by field work and data gathering, data analysis and development of the final output—the CHI map—for the study area. It allows as input data, data gathered by traditional methods (vantage points observations) as well as more advanced methodologies such as RADAR, as long as the data is present in the form of avian routes.

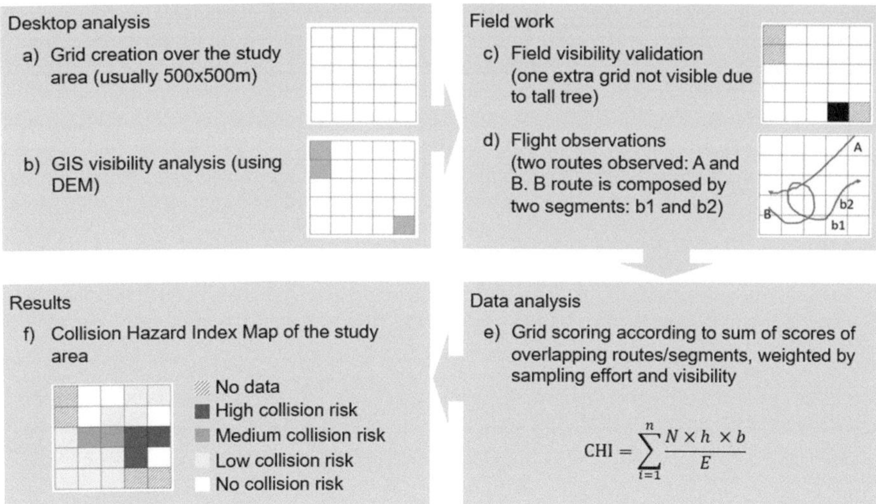

Fig. 4.1 Methodological steps for the collision hazard index (CHI) analysis, grouped in four main categories: **Desktop analysis**—the step included in this group involves the creation of the 500 × 500 m grid over all the wind farm study area and a preliminary visibility analysis (step *a* and *b*); **Field work**—where the preliminary visibility analysis of the area is validated and avian routes and behaviour are registered; **Data analysis**—scores are given to each route segment according to the observed behaviour and height categories. Grid scoring is then calculated by adding the scores of all overlapping route segments and weighted by the sampling effort and degree of visibility; **Results**—a grid map of the study area is created presenting the resulting scores of the index: grids with higher levels of CHI are considered areas where avian species are at higher risk of collision

Table 4.2 Flight behaviour categories. Description of types of avian flight behaviour that must be registered during field work and further analysed during the collision hazard index (CHI) estimation

Category name	Description
Circular flights	Includes the soaring flights, common in birds of prey as a way of flying long distances and gain altitude with low energy expense, taking advantage of pressure differences between different temperature air masses
Hunting flights	Includes hunting flight, hovering and nose-diving
Territorial flights	Exhibition or chasing flights, usually frequents near nesting sites
Passing flights	Includes the gliding and directional flights that don't fit in any of the other categories
Resting	Flight is absent. This behaviour is usually observed in utility poles, trees and rocks. While it is often associated with resting and grooming, it is also associated with a hunting method used by some species, like Kestrels. Under unfavourable conditions, it is sometimes difficult to distinguish between the two situations (Village 1990)

Table 4.3 Flight height categories. Description of the levels of flight height that must be registered during field work and further analysed during the collision hazard index (CHI) estimation

Category name	Description
Under the rotor	Flights seen or estimated to be under the sweeping area of the present or future rotor (usually between 0 and 40 m of height, in Portugal[a])
Rotor height	Flights seen or estimated to be under the sweeping area of the present or future rotor (usually between 40 and 120 m of height, in Portugal[a])
Above the rotor	Flights seen or estimated to be under the sweeping area of the present or future rotor (usually above 120 m of height, in Portugal[a])

[a]This height limit depends on the characteristics of the turbines of a given wind farm, namely hub height and rotor diameter

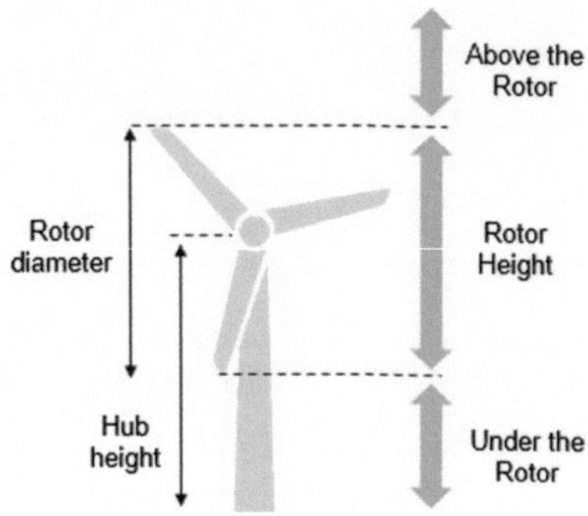

Fig. 4.2 Flight height categories considered for the collision hazard index (CHI) analysis. Categories are determined in relation to the wind turbine's sweeping area and height given by the rotor's diameter and hub height

Field Methodology

Traditional data collection is usually obtained through point counts (vantage points) that cover the wind farm and the surrounding area, but data gathering can also be conducted with data collected by telemetry or RADAR. Despite the chosen field collection method, birds' movements must be mapped (Fig. 4.1, step d; Fig. 4.4a), and described with as much detail as possible regarding the bird's height of flight and flight behaviour and in accordance to the categories presented in Tables 4.2 and 4.3. Note that the flight height categories are determined in relation to the present or future wind turbine's sweeping area and height (Fig. 4.2).

In the traditional data collection methodology, field observers write down on a map of the area the routes of all avian species crossing the area, giving codes to the routes so that they can be cross-referenced to the species, number of individuals, behaviour and height categories. Every time the observed individual(s) on a route (route A in Fig. 4.4a for example) changes its behaviour or height category, the route is divided into segments and a sub-reference is given (route A is divided into segment a1 to a5) so that each segment can be cross-referenced.

Table 4.4 Scores used in the collision hazard index (CHI) calculations

Scores	Flight behaviour category	Flight height category
Score 1	All other flight categories	Above the rotor
Score 2	Circular flights, hunting, hovering and territorial flight	Under the rotor
Score 4	No categories	At rotor height

Each avian route must be separated in segments according to the observed flight behaviour and height categories, allowing further calculation of a sum of scores for each segment

Desktop Analysis, Field Data Analysis and Results

The gathered data is inserted in a GIS tool and analysed based on a regular grid previously prepared during desktop analysis phase (Fig. 4.1, step a; Fig. 4.4c). The grid is a usually a 250 or 500 m^2 grid (depending on the size of the wind farm area) and therefore is the basis of all of the following analysis.

To translate collision risk into a value, numerical scores are associated to different flight behaviours and flight height categories. Since each route is divided into segments according to changes in categories of behaviour and height, each segment of bird route is given a scoring, according to the following (Table 4.4).

Collision risk is then summed for each square grid, by adding the scores of all route segments observed in the grid's area (Fig. 4.1, step e) using the following CHI formula (Eq. 4.1):

$$\mathrm{CHI} = \sum_{i=1}^{n} \frac{N \times h \times b}{E} \tag{4.1}$$

where,

- n represents the number of bird route segments in each square grid;
- N the number of individuals in each route;
- h the score attributed for the height of the bird at each segment of the route;
- b the score attributed to the flight behaviour of the bird at each segment of the route;
- E the sampling effort in each square grid, since each grid may have different efforts if, for example, sampled from different sampling points.

However, since not all squares of the grid in the study area are equally surveyed (due to range limitation of the observer/telemetry/RADAR, presence of vegetation that might reduce the ability to see/detect or because of sampling locations that overlap in visibility/detection range), a visibility/detection analysis must be performed for each square so as to weight the CHI result for each square, both in terms of number of hours of observation and by degree of visibility/detection (sampling effort, *E*).

For the visibility analysis, areas that are effectively observable/detectable from each of the vantage points are obtained by means of a GIS analysis, undertaken by crossing the observers' elevation from a Digital Elevation Model (DEM with 100 m

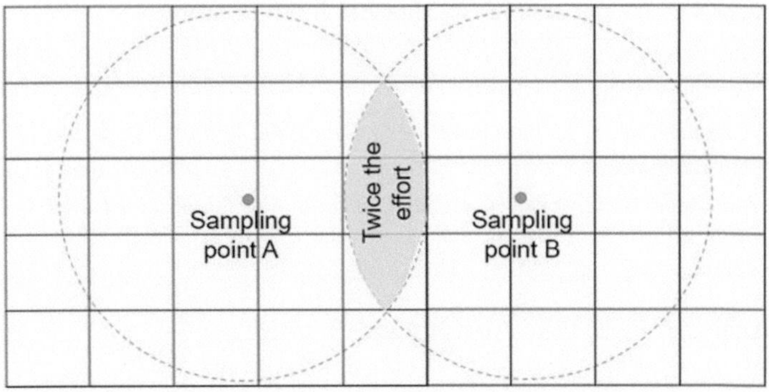

Fig. 4.3 Example of field work sampling effort estimation. Due to the overlapping visibility range of two sampling points, some square grids of the study area have higher sampling efforts and have to be adjusted when applying the collision hazard index (CHI) analysis

resolution) with the height of flights that might surpass obstacles. For this, an algorithm is generated so as to simulate birds flying above the ground, which adds a determined value (usually 20–50 m) to each DEM cell's height in the moment their visibility/detection is being evaluated. The visibility/detection analysis algorithm results in a final output grid map where visible and not visible areas (or detectable/undetectable areas if the chosen detection method is telemetry or RADAR) (Fig. 4.1, step b). Since the DEM has low resolution (relatively big square cells) and the presence of vegetation might represent an additional obstacle in terms of visibility, the final resulting visibility map is then validated by a technician while in the field (Fig. 4.1, step c).

In some cases, the visibility range areas of each sampling point might overlap and some squares will be observed from both sampling points during the surveys, therefore, the sampling effort for those squares will be higher than in squares where there is no overlap (Fig. 4.3). In this case, the total number of bird observation for each species must be weighted by the total number of observation hours at each square.

Finally, the resulting scores for each grid square are mapped and a colour scale is used to represent areas with higher number of observed flights that are considered to contribute to higher risk of collision (Fig. 4.1, step f; Fig. 4.4d). Therefore, the CHI map reflects the horizontal and the vertical use of the area in terms of number of risky behaviours exhibited by the target species.

The results of CHI analysis allows areas of higher avian collision risk to be identified. Therefore, turbines within these areas are expected to cause higher fatalities during the operation phase (refer to the next subchapter). This type of analysis can be used for turbine layout adjustments during EIAs in the pre-construction phase (Marques et al. 2010), where turbines located in areas of higher collision risk should be recommended to be moved to adjacent areas of lower risk, if feasible. If this is not possible, then mitigation measures can be recommended in order to minimise impact during the operation phase (Marques et al. 2010).

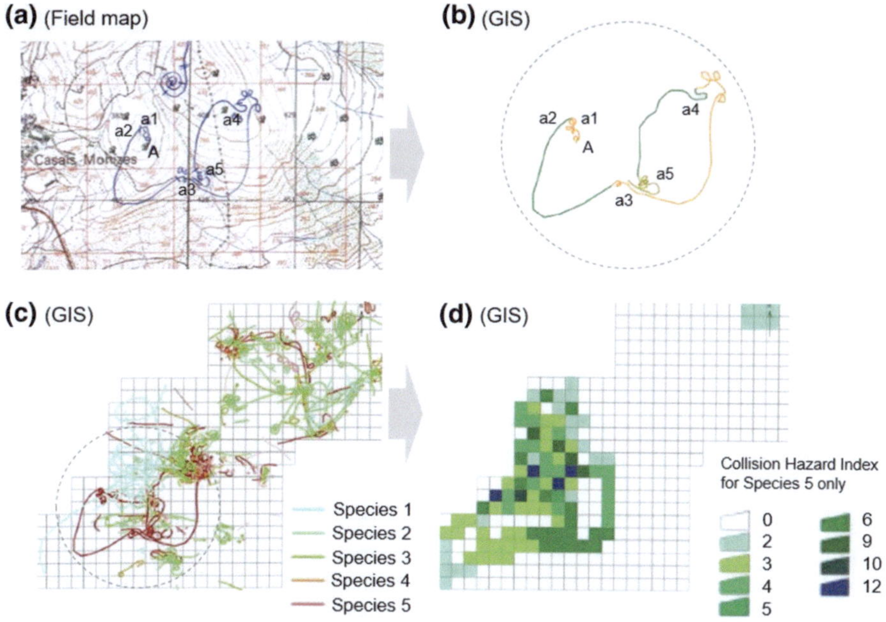

Fig. 4.4 Example output of each step of the collision hazard index (CHI) analysis: **a** mapping avian movements in the field during vantage points. Topographic maps are used for reference and aid the observer in drawing the avian flying routes; **b** detail of a mapped movement of a bird (route) after being introduced into a GIS tool: each colour of the line represents changes in the bird's behaviour and corresponds to a 'route segment'; **c** GIS plotting of all birds' movements in the area; **d** CHI map of a Portuguese wind farm (darker colours represent areas of higher avian risk of target species)

Although this analysis was established with the aim of predicting the impacts of a wind farm during the pre-construction phase, this approach can and has also been widely used in Portugal for projects in operational phases (refer to examples: Bio3 2009b, 2010, 2011, 2012, 2013, 2014b; Cordeiro et al. 2013; Bioinsight 2016) to assess differences in space use (both vertical and horizontal) throughout post-construction monitoring years. Changes in space use along the years, if analysed in parallel with distance to turbines and mortality, can be an indicator of a displacement effect (species avoiding the area) or an indicator of consequences due to mortality (Marques et al. 2010).

4.5.1.2 The Common Kestrel Example at a Portuguese Wind Farm

Candeeiros wind farm is located in a ridge of a limestone mountain in central Portugal, inside a Protected Area (Natural Park "Serra de Aire e Candeeiros") and a Natura 2000 Site (PTCON0015). The wind farm consists of 37 turbines located in

Fig. 4.5 Common kestrel (*F. tinnunculus*) found at Candeeiros wind farm during a fatality search survey

an area (ca. 70 km^2) of calcareous hills with lots of quarries and cliffs used by common kestrels (*F. tinnunculus*; listed in Portugal as a least concern species; Cabral et al. 2006) for nesting (Cordeiro et al. 2013). The wind farm became operational in 2005 and has been monitored since then. The wind farm presents high mortality of the species at certain turbines, with 18 registered fatality events in the 11 years of operation (see Bioinsight 2016) (Fig. 4.5).

The monitoring consists of general bird surveys, including point counts for passerines and raptors, and weekly fatality searches, by humans, around all wind turbines (50 m radius around each turbine). Raptor-dedicated surveys involve 1 h point counts, where the number of individuals and the flight behaviour are monitored. After the first results of mortality of the species (in year one of operation phase), the monitoring reports of the wind farm included a CHI for the species, and, in 2016, a time series of the index was presented, reflecting areas of higher risk across the wind farm area along the years. In a study presented by Cordeiro et al. (2013), a time series of CHI maps and cumulative fatalities of kestrels since 2005 was also presented and it was possible to see that the turbines where higher levels of kestrel fatalities were observed over the years were coincident with the areas where higher levels of risk of collision were estimated. Figure 4.6 shows one of these CHI maps from the year 2012, overlapped with cumulative kestrel fatalities (2015–2012). These results suggest that higher fatalities were observed in turbines coincident with the areas where the CHI analysis showed higher levels which in turn are also the areas where the kestrels exhibited higher risk behaviours (such as higher number of flights at height of the sweeping area).

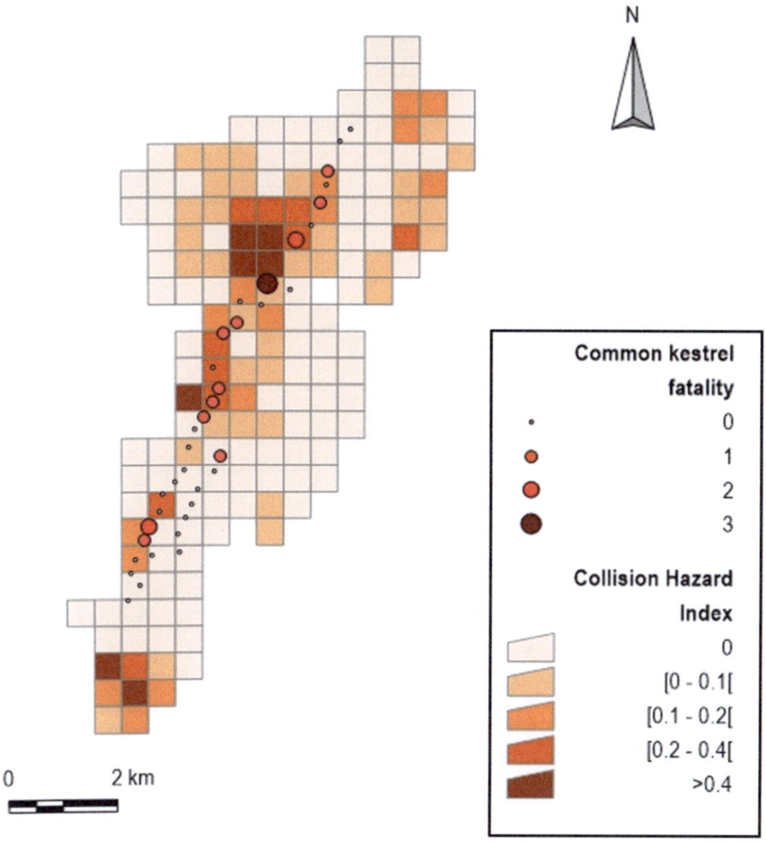

Fig. 4.6 Collision risk analysis for the common kestrel (*F. tinnunculus*) at Candeeiros wind farm during 2012 and cumulative fatalities of the same species since 2005 (adapted from Cordeiro et al. 2013). Areas presenting higher levels of collision hazard index (CHI) are coincident with turbines where higher fatalities occurred, suggesting a relation between the CHI analysis and the estimation of turbines were kestrels are prone to collision

References

Band, W., Madders, M., & Whitfield, D. P. (2007). In M. de Lucas, G. F. E. Janss, & M. Ferrer (Eds.), *Developing field and analytical methods to assess avian collision risk at wind farms.* Madrid: Quercus.

Band, B. (2012a). Using a collision risk model to assess bird collision risks for offshore windfarms —with extended method: Worked example. SOSS Report. Retrieved from www.bto.org

Band, B. (2012b). Using a collision risk model to assess bird collision risks for offshore windfarms. SOSS Report. The Crown Estate.

Barrios, L., & Rodríguez, A. (2004). Behavioural and environmental correlates of soaring-bird mortality at on-shore wind turbines. *Journal of Applied Ecology, 41*(1), 72–81.

Beston, J. A., Diffendorfer, J. E., Loss, S. R., & Johnson, D. H. (2016). Prioritizing avian species for their risk of population-level consequences from wind energy development. *PLoS ONE, 11* (3), e0150813. doi:10.1371/journal.pone.0150813

Bio3. (2008). Parque Eólico da Serra do Mú - Monitorização da comunidade de aves e quirópteros. Relatório final (Fase 1 – anterior à construção). Report compiled by Bio3 - Estudos e Projectos em Biologia e Valorização de Recursos Naturais, Lda. Almada, February 2008. 69 p (in Portuguese).

Bio3. (2009a). Parque Eólico de Malhanito - Monitorização da comunidade de aves e quirópteros. Relatório 1 (Fase 1 – anterior à construção). Report compiled by Bio3 - Estudos e Projectos em Biologia e Valorização de Recursos Naturais, Lda. Almada, July 2009. 55 p (in Portuguese).

Bio3. (2009b). Parque Eólico da Serra do Mú. Monitorização da comunidade de aves e quirópteros. Relatório final (Fase I1 – construção). Report compiled by Bio3 - Estudos e Projectos em Biologia e Valorização de Recursos Naturais, Lda.. Almada, January 2009. 64 p (in Portuguese).

Bio3. (2010). Parque Eólico da Lousã II. Monitorização da comunidade de aves. Relatório III (Fase III – Exploração). Report compiled by Bio3 - Estudos e Projectos em Biologia e Valorização de Recursos Naturais, Lda.. Almada, November 2010. 63 p (in Portuguese).

Bio3. (2011). Sub-Parque Eólico Montalegre. Monitorização das comunidades de aves e quirópteros. Relatório 3 (Fase 3 – exploração). Report compiled by Bio3 - Estudos e Projectos em Biologia e Valorização de Recursos Naturais, Lda. Almada, September 2011. 145 p (in Portuguese).

Bio3. (2012). Parque Eólico do Carreço-Outeiro II. Monitorização da comunidade de aves e quirópteros. Relatório III (Fase III - exploração). Report compiled by Bio3 - Estudos e Projectos em Biologia e Valorização de Recursos Naturais, Lda. Almada, October 2012. 127 pp (in Portuguese).

Bio3. (2013). Parque Eólico da Terra Fria. Monitorização das Comunidades de Aves e Quirópteros. Relatório V (Fase de exploração Montalegre). Relatório Final Contim. Report compiled by Bio3 - Estudos e Projectos em Biologia e Valorização de Recursos Naturais, Lda. Almada, April 2013. 224 p (in Portuguese).

Bio3. (2014a). Parque Eólico de Três Marcos. Monitorização da comunidade de Aves. Relatório I (Fase anterior à construção – Ano 2012/2013). Report compiled by Bio3 - Estudos e Projectos em Biologia e Valorização de Recursos Naturais, Lda.. Almada, Jully 2014. 65 p (in Portuguese).

Bio3. (2014b). Parque Eólico do Malhanito. Monitorização das Comunidades de Aves e Quirópteros. Relatório III (Fase de exploração – 2012/2013). Report compiled by Bio3 - Estudos e Projectos em Biologia e Valorização de Recursos Naturais, Lda.. Almada, March 2014. 178 p (in Portuguese).

Bioinsight. (2016). Parque Eólico da Serra de Candeeiros. Monitorização da comunidade de aves. Relatório 11 (Fase de exploração – ano 2015). Report compiled by Bioinsight, Lda. Odivelas, November 2016 (in Portuguese).

Bispo, R., Marques, A. T., Soares, M., Costa, H., Mascarenhas, M., & Bernardino, J. (2015). *Using bird fatality explanatory variables and conservation status to identify primary target species*. Paper presented at the Conference on Wind Energy and Wildlife Impacts, Berlin, March 10–12.

Bolker, E. D., Hatch, J. J., & Zara, C. (2014). Modeling how windfarm geometry affects bird mortality. arXiv:1408.1580 (preprint).

BTO. (2005). Appraisal of Scottish Natural Heritage's wind farm collision risk model and its application. A report by the British Trust for Ornithology under contract to English Nature, BTO Research Report 401. Thetford, Norfolk.

Cabral, M. J. (coord.), Almeida, J., Almeida, P. R., Dellinger, T., Ferrand de Almeida, N., & Oliveira, M. E. et al. (Eds.) (2006). Livro Vermelho dos Vertebrados de Portugal 2ª ed. Instituto da Conservação da Natureza/Assírio & Alvim. Lisboa (in Portuguese).

Carrete, M., Sánchez-Zapata, J. A., Benítez, J. R., Lobón, M., & Donázar, J. A. (2009). Large scale risk-assessment of wind-farms on population viability of a globally endangered long-lived raptor. *Biological Conservation, 142,* 2954–2961.

Chamberlain, D. E., Rehfisch, M. R., Fox, A. D., Desholm, M., & Anthony, S. J. (2006). The effect of avoidance rates on bird mortality predictions made by wind turbine collision. *Ibis, 148,* 198–202.

Cook, A. S. C. P., Humphreys, E.M., Masden, E.A., Burton, N.H.K. (2014). The Avoidance Rates of Collision Between Birds and Offshore Turbines. Scottish Marine and Freshwater Science, Vol. 5, Number 16. Published by Marine Scotland Science. ISSN: 2043-7722.

Cordeiro, A., Mascarenhas, M., & Costa, H. (2013). *Long term survey of wind farms impacts on Common Kestrel's populations and definition of an appropriate mitigation plan.* Paper presented in the Conference in Wind Power and Environmental Impacts, Stockholm, February 5–7.

de Lucas, M., Janss, G. F. E., & Ferrer, M. (2008). Collision fatality of raptors in wind farms does not depend on raptor abundance. *Journal of Applied Ecology, 45,* 1695–1703.

de Lucas, M., Ferrer, M., Janss, G. F. E., & Magar, V. (2012) Using wind tunnels to predict bird mortality in wind farms: The case of Griffon Vultures. *PLoS ONE, 7*(11), e48092

Desholm, M. (2006). Wind farm related mortality among avian migrants—a remote sensing study and model analysis. *Ph.D. Dissertation.*

Drewitt, A. L., & Langston, R. H. W. (2006). Assessing the impacts of wind farms on birds. *Ibis, 148,* 29–42.

Eichhorn, M., Johst, K., Seppelt, R., & Drechsler, M. (2012). Model-based estimation of collision risks of predatory birds with wind turbines. Ecology Society 17.

FAU & LSE (n.d.). Bestimmung des Kollisionsrisikos von Fledermäusen an Onshore-Windenergieanlagen in der Planungspraxis. Retrieved December, 2016, form www.windbat.techfak.fau.de (in German).

Furness, R. W., Wade, H. M., & Masden, E. A. (2013). Assessing vulnerability of marine bird populations to offshore wind farms. *Journal of Environmental Management, 119,* 56–66.

Garthe, S., & Hüppop, O. (2004). Scaling possible adverse effects of marine wind farms on seabirds: Developing and applying a vulnerability index. *Journal of Applied Ecology, 41*(4), 724–734.

Hanssen, F., & May, R. (2015). *Avian sensitivity mapping and wind energy: GIS-based wind turbine micrositing tool.* Paper presented at the Webinar #4 in WREN Environmental Webinar Serie.

Holmstrom, L., Hamer, T., Colclazier, E., Denis, N., Verschuyl, J. P., & Ruche, D. (2011). Assessing avian–wind turbine collision risk: An approach angle dependent model. *Wind Engineering, 35,* 289–312.

Johnson, G. D., Erickson, W. P., Strickland, M. D., Shepherd, M. F., Shepherd, D. A., & Sarappo, S. A. (2002). Collision mortality of local and migrant birds at a large-scale windpower development on Buffalo Ridge, Minnesota. *Wildlife Society Bulletin, 30,* 879–887.

Johnston, A., Cook, A. S. C. P., Wright, L. J., Humphreys, E. M., & Burton, N. H. K. (2014). Modelling flight heights of marine birds to more accurately assess collision risk with offshore wind turbines. *Journal of Applied Ecology, 51,* 31–41.

Katzner, T. E., Brandes, D., Miller, T., Lanzone, M., Maisonneuve, C., Tremblay, J. A., et al. (2012). Topography drives migratory flight altitude of golden eagles: Implications for on-shore wind energy development. *Journal of Applied Ecology, 49*(5), 1178–1186.

Kerlinger, P. (1989). *Flight strategies of migrating hawks.* University of Chicago Press.

Krijgsveld, K. L., Akershoek, K., Schenk, F., Dijk, F., & Dirksen, S. (2009). Collision risk of birds with modern large wind turbines. *Ardea, 97*(3), 357–366.

Marques, A. T., Bernardino, J., Costa, M., & Mascarenhas, M. (2010). Following raptors and soaring birds populations in wind farms: A methodological protocol. European Bird Census Council—EBCC. Cáceres, March 22–26, 2010.

Marques, A. T., Batalha, H., Rodrigues, S., Costa, H., Pereira, M. J. R. P., Fonseca, C., et al. (2014). Understanding bird collisions at wind farms: An updated review on the causes and possible mitigation strategies. *Biological Conservation, 179,* 40–52.

Masden, E., & Cook, A. (2016). Avian collision risk models for wind energy impact assessments. *Environmental Impact Assessment Review, 56,* 43–49.

McAdam, B. J. (2005). A Monte-Carlo model for bird/wind turbine collisions. *M.Sc. Dissertation*. Ecology University of Aberdeen.

Newton I. (2007). *The migration ecology of birds*. Academic Press. 984 p.

Noguera, J. C., Pérez, I., & Mínguez, E. (2010). Impact of terrestrial wind farms on diurnal raptors: Developing a spatial vulnerability index and potential vulnerability maps. *Ardeola, 57*(1), 41–53.

Normandeau Associates. (2012). A habitat-based wind-wildlife collision model with application to the upper great plains region. Final Report for the U.S. Department of Energy. Gainesville, Florida, 98 p.

Percival, S. M. (2005). Birds and wind farms: What the real issues? *British Birds, 98,* 194–204.

Podolsky, R. (2008). Method of and article of manufacture for determining probability of avian collision. US Patent 7,315,799, 1.

Schaub, M. (2012). Spatial distribution of wind turbines is crucial for the survival of red kite populations. *Biological Conservation, 155,* 111–118.

SNH. (2000). Windfarms and birds: Calculating a theoretical collision risk assuming no avoiding action. Report compiled by Scottish Natural Heritage. Guidance Note Series.

SNH. (2009). Collision avoidance of golden eagles at wind farms under the 'Band' Collision risk model. Report from Natural Research to Scottish Natural Heritage (SNH). Banchory, UK. Retrieved December 2016, from www.snh.org.uk/pdfs/strategy/renewables/b362718.pdf

Smales, I., Muir, S., Meredith, C., & Baird, R. (2013). A description of the Biosis model to assess risk of bird collisions with wind turbines. *Wildlife Society Bulletin, 37,* 59–65.

Soares, M. J. S. (2014). An avian relative fatality risk index for iberian species on wind farms based on zero inflated count models. *M.Sc. dissertation*, University of Aveiro, Portugal.

SPEA. (2005). *A energia eólica e a conservação da avifauna em Portugal*. Lisbon, Portugal: Sociedade Portuguesa para o Estudo das Aves (in Portuguese).

Tellería, J. L. (2008). Wind power plants and the conservation of birds and bats in Spain: A geographical assessment. *Biodiversity and Conservation, 18*(7), 1781–1791.

Tham, M. (2000). *Overview of mechanistic modeling techniques*. Retrieved September 20, 2012, from http://lorien.ncl.ac.uk/ming/dynamics/modelling.pdf

Tucker, V. A. (1996a). A mathematical model of bird collisions with wind turbine rotors. *Journal of Solar Energy Engineering* 253–262.

Tucker, V. A. (1996b). Using a collision model to design safer wind turbine rotors for birds. *Journal of Solar Energy Engineering, 118,* 263–269.

UHI. (2015). Developing an avian collision risk model to incorporate variability and uncertainty. Report by University of the Highlands and Islands (UHI).

USFWS. (2013). Eagle conservation plan guidance module 1—land-based wind energy. Report developed by the U.S. Fish and Wildlife Service.

Village, A. (1990). *The kestrel*. London: A&C Black Publishers Ltd.

Willmott, J., Costello, E. A., Gordon, C., Forcey, G., Casto S., & Beaulac, G., et al. (2012). Bird and Bat Collision Risks & Wind Energy Facilities. Inter-American Development Bank. Environmental Safeguards Unit. Discussion paper No. IDB-DP-354.

WindBat. (2016). *WindBat web site*. Retrieved December 2016, from www.windbat.techfak.fau.de

Chapter 5
The Indirect Impacts of Wind Farms on Terrestrial Mammals: Insights from the Disturbance and Exclusion Effects on Wolves (*Canis lupus*)

Gonçalo Ferrão da Costa, João Paula, Francisco Petrucci-Fonseca and Francisco Álvares

5.1 General Overview

In a growing number of countries, including Portugal, wind power is becoming increasingly important as a power supply. According to the Portuguese Renewable Energy Association (APREN), presently, one quarter of the consumed energy in Portugal comes from wind (APREN 2017). Due to the functional characteristics of wind turbines, most of the impact studies have focused on flying vertebrates, specifically birds and bats, by addressing activity patterns, habitat use and mortality (e.g., Baerwald and Barclay 2009; Drewitt and Langston 2006; Kunz et al. 2007; Pearce-Higgins et al. 2009). Impacts on non-flying vertebrates have been poorly studied, despite several findings that terrestrial animals can be affected by wind power development in various ways (de Lucas et al. 2005; Rabin et al. 2006; Walter et al. 2006; Santos et al. 2010; Álvares et al. 2011; Lovich and Ennen 2013; Agha et al. 2015; Lopucki and Mros 2016).

G. Ferrão da Costa (✉) · J. Paula
Bioinsight, Odivelas, Portugal
e-mail: goncalo.c@bioinsight.pt

J. Paula
e-mail: j.paula@bioinsight.pt

F. Petrucci-Fonseca
Grupo Lobo, Department of Animal Biology and CE3C—Centre for Ecology, Evolution and Environmental Changes, Faculty of Sciences, University of Lisbon, Lisbon, Portugal
e-mail: fpfonseca@fc.ul.pt

F. Álvares
CIBIO/InBIO—Research Center in Biodiversity and Genetic Resources, University of Porto, Vairão, Portugal
e-mail: falvares@cibio.up.pt

M. Mascarenhas et al. (eds.), *Biodiversity and Wind Farms in Portugal*,
https://doi.org/10.1007/978-3-319-60351-3_5

Habitat modifications and behavioural changes are described as impacts that can occur on terrestrial vertebrates, due to several actions inherent to wind farms construction and operation: turbines installation, roads construction, transformers, substations and power lines (Kuvlesky et al. 2007). All these infrastructures may cause direct and permanent habitat loss and fragmentation, even though this might be of minor importance for large terrestrial mammals with extensive home ranges (Helldin et al. 2012). The increased human presence related to wind farm construction and operation can also augment wildlife disturbance, particularly during sensitive periods, such as breeding season, and might also increase the risk of mortality by collisions with vehicles along access roads (Kuvlesky et al. 2007; Northrup and Wittemyer 2013). According to a review by Lovich and Ennen (2013), the construction and operation of wind farms have both potential and known impacts on terrestrial vertebrates, such as: (i) increase in direct mortality due to traffic collisions; (ii) destruction and modification of the habitat, including road development, habitat fragmentation and barriers to gene flow; (iii) noise effects, visual impacts, vibration and shadow flicker effects from turbines; (iv) electromagnetic field generation; (v) macro and microclimate change; (vi) predator attraction; and (vii) increase in fire risks.

Regarding the effect of wind farms on large mammals, Helldin et al. (2012) concluded that these vertebrates are affected by the increase of human activity during construction and operational phases. According to these authors, large animals may temporarily avoid wind farms during the construction phase, but when machinery and human presence decrease, animals seem to acclimatise to wind farms, although these responses may vary with species, sex, age, individual, time of the year or type of disturbance. Helldin et al. (2012) also highlighted that the development of road networks associated with wind farms could promote increased access for traffic related to recreation, forestry, agriculture and hunting. The consequence, particularly on remote places, is the increase in human presence, affecting large mammals via significant disturbance, habitat loss and habitat fragmentation. These negative effects are expected to be particularly relevant for species that are more sensitive to human presence and activities, such as large carnivores.

Large carnivores, such as the wolf, bear, lynx or wolverine, tend to avoid areas that are regularly used by humans and—especially for breeding—show a preference for rugged and undisturbed areas (Theuerkauf et al. 2003; George and Crooks 2006; May et al. 2006; Elfstrom et al. 2008; Sazatornil et al. 2016), which are often chosen for wind power development (Passoni et al. 2017) . However, despite these taxa being the regular focus of research and public interest, currently, there is scant knowledge about wind farms' effects on large carnivores. The only known studies on large carnivores were focused on black bears (Wallin 1998), wolverines (Flagstad and Tovmo 2010), and Portuguese wolves (Álvares et al. 2011, 2017), all reporting some avoidance during the construction phase of wind farms. The lack of robust studies either presenting results of a long term BACI (before-after-control-impact) monitoring programme or enough solid data to rule out the effect of other factors, renders the need for further research focusing on the effects of wind farms on large carnivores, particularly for species with a wide range such as wolves.

Impacts of wind farms on wolves may be inferred from the available knowledge about wind farms effects on other large mammals or, instead, from other human activities on wolves (Helldin et al. 2012). In fact, construction and operation of a wind farm involves a set of actions that is very similar to those associated with other infrastructures or human activities such as road development or forest logging. The effects of extensively logged areas and high density of roads on wolves have been widely studied (Person and Russell 2009; Houle et al. 2010; Taylor 2010; Zimmermann et al. 2014) and can provide us insights on the complexities of behavioural responses of wolves to wind farms.

The density of roads (either gravel or paved roads) and logging can have profound effects on wolf movements, habitat use and den site selection, but may also be associated with ambivalent responses from wolves depending on several factors, such as site characteristics and cumulative effects of other sources of disturbance (Zimmermann et al. 2014). Earlier studies conducted in the USA (Thiel 1985; Mech et al. 1988) suggested low road density thresholds (e.g., <0.58 km/km^2) may be necessary for wolves to endure, although Fuller et al. (1992) pointed that this threshold could be slighty higher (0. 7 km/km^2) in areas with lower human density. However, recent studies in Europe highlighted the adaptive capabilities of wolves, with the species found to be present in areas with average road densities as high as 1.2 km/km^2 (Theuerkauf et al. 2003; Jedrzejewski et al. 2005; Llaneza et al. 2012), reaching even 1.9 km/km^2 in some areas of the Iberian Peninsula (Dennehy et al. 2013). Wolf responses to road accessibilities are also dependent on several cumulative factors, such as habitat quality, human density and prey availability (Eggermann et al. 2011; Llaneza et al. 2012). Furthermore, the expansion of new traffic accessibilities can cause an increase in human-related mortality on wolves (Zimmermann et al. 2014), presenting a significant negative impact particularly during sensitive periods such as the breeding season. On the other hand, wolves may benefit from the environmental changes induced by wind farms, by preferably using these habitat discontinuities as areas of higher prey availability (Berger 2007). This evidence reinforces the need for local monitoring programmes so that the specific interactions between wolves and wind farms can be fully understood.

In this chapter, based on monitoring plans that have been conducted by the authors, we aimed to summarise the available knowledge about the effects of wind farms on wolves in Portugal, along with insights about methodological approaches, and mitigation and compensation measures.

5.2 Wolves and Wind Farms in Portugal

5.2.1 Setting the Context: Wolf Distribution, Landscape Specificities and Environmental Issues

Until the beginning of the twentieth century, wolves were found across all Portuguese continental territory, but, over a few decades, this species became

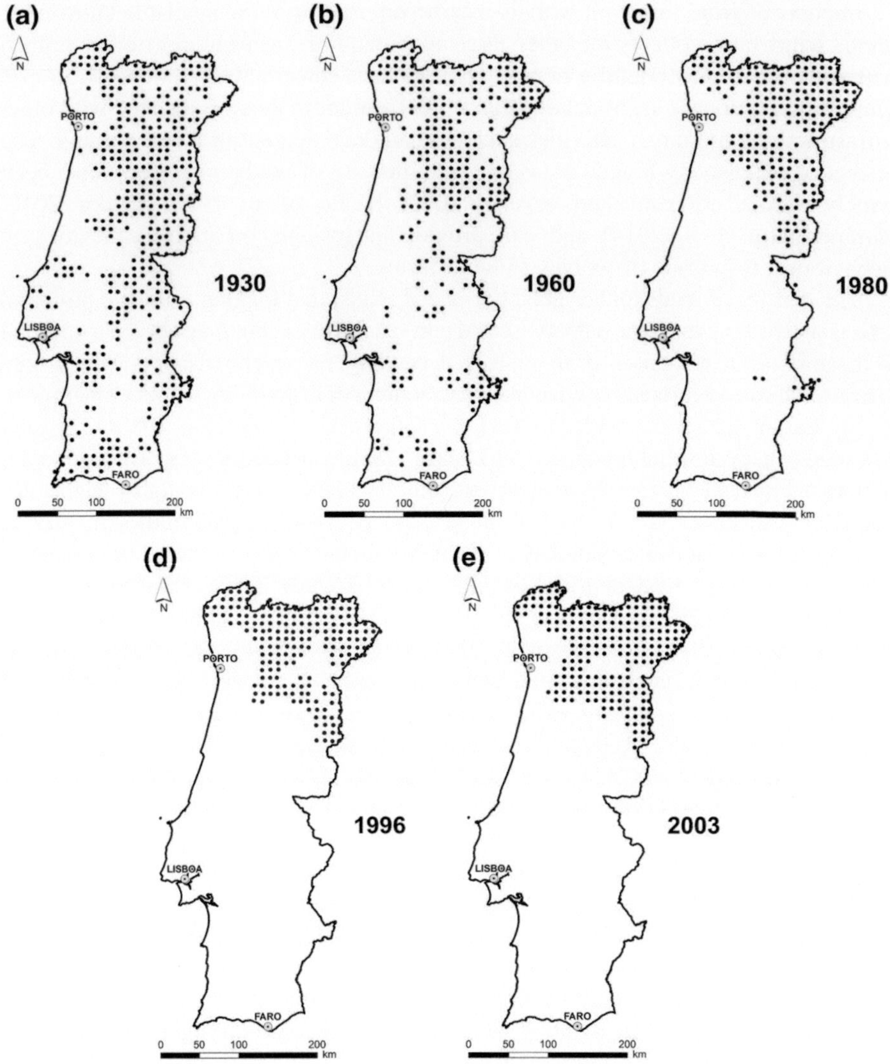

Fig. 5.1 Wolf distribution in Portugal over the last century: **a** 1930, **b** 1960, **c** 1980, **d** 1996, **e** 2003. Adapted from Petrucci-Fonseca (1990) and Pimenta et al. (2005)

extinct from the southern and coastal areas, persisting only in the northern mountainous regions (Petrucci-Fonseca 1990; Fig. 5.1a–e). This sharp decline was probably due to a combination of factors such as a decrease in food availability (both wild and domestic prey), habitat fragmentation from expanding human activities and increasing human persecution as a result of widespread access to firearms and poison (Petrucci-Fonseca 1990).

This population decline induced Portuguese authorities to impose a restrictive legal framework to protect the wolf. Since 1988, this species has been fully protected by national law, which forbids the killing or capturing of individuals or disturbing important areas, such as breeding sites. Additionally, and according to European Community legislation (Habitats Directive 92/43/EEC) the wolf is considered a priority species for conservation, included in the Annexes II and IV, where Sites of Community Importance (SCIs) should be created and managed in accordance with the ecological needs of the species. The wolf is classified as Endangered (EN) by the Portuguese Vertebrate Red Data Book (Cabral et al. 2006) and approximately 30% of its range is covered by National Protected Areas or SCIs.

Since the late 1990s, the wolf range in Portugal has stabilised at approximately 20,000 km^2, corresponding only to 20% of its original distribution area (Petrucci-Fonseca 1990). Currently, and according to the last Portuguese wolf census (Pimenta et al. 2005), the wolf population is estimated at between 200 and 400 individuals (approximately 65 breeding packs), comprising two sub-populations divided by the Douro river (Fig. 5.2). The northern population has between 45 and 54 packs and is continuous with the larger Spanish wolf population, estimated at around 2000 individuals (Álvares et al. 2005) (Fig. 5.2). Nonetheless,

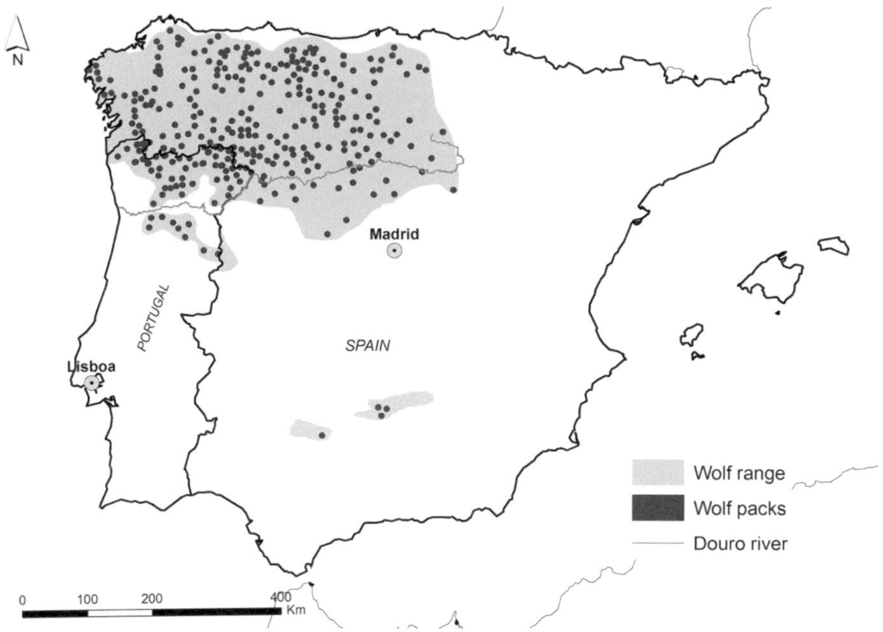

Fig. 5.2 Wolf distribution and location of known packs in the Iberian Peninsula. Adapted from Álvares et al. (2005)

a population decline in some of these northern areas has been predicted if current trends in land use changes, livestock abundance and road density continue to operate (Santos et al. 2007). The portuguese wolf population located just south of Douro river is further reduced, with only six to nine packs, exhibiting a low reproduction rate and a high degree of fragmentation and isolation as suggested by genetic studies (Godinho et al. 2007), which turns it in one of the fewest wolf populations in Europe considered at risk of extinction (Boitani and Ciucci 2009).

Although wolves are often considered a symbol of wilderness and remote places (Mech and Boitani 2003), in the Iberian Peninsula and particularly in Portugal, this carnivore has been living in a very humanised landscape for several centuries, and has adapted to human presence, activities and persecution (Álvares et al. 2011; Eggermann et al. 2011, Llaneza et al. 2012). Wolves can live wherever they have enough food and are not killed by humans (Mech and Boitani 2003), and in Portugal those conditions mostly occur in the mountainous areas of the North and Centre (between 800 m and 1500 m above sea level), which became their stronghold due to lower human density, difficult accessibility and higher prey availability (Grilo et al. 2002; Eggermann et al. 2011). At these domains, the landscape is dominated by rocky scrublands, with scarce forest cover due to livestock grazing and wild fires. The scarcity of forest and woodland results in few available wild prey: roe deer (*Capreolus capreolus*), while expanding, still have low densities (Torres et al. 2015a); red deer (*Cervus elaphus*) are very localised within wolf ranges or occur at low densities (Vingada et al. 2010); while wild boar (*Sus scrofa*) are the only prey with wide occurrence and high densities in recent years (Vingada et al. 2010). This ecological context led wolves to feed mainly on domestic prey, such as cattle, goats, sheep and horses, which comprise up to 80% of the wolf diet in most of Portugal (Vos 2000; Álvares et al. 2015; Torres et al. 2015b). Consequently, there is a big conflict between livestock owners and this predator, resulting in high wolf mortality due to human causes (illegal hunting, poisoning and trapping), despite legal protection and considerable conservation efforts (Álvares et al. 2015).

The recent development of wind farms in these mountainous areas within the wolf range posed a challenge for wolf conservation, mainly due to the increased human accessibility and the consequent disturbance to areas that were previously quite remote.

5.2.2 Methodological Approach

Considering the current endangered and legal status of wolves in Portugal, the national authorities imposed the measure that every new infrastructure (e.g., wind farm, road, dam) projected within the wolf range should be subject to an Environmental Impact Assessment (EIA) study, to evaluate and minimise potential negative effects on this carnivore. If negative impacts are expected, the project's approval is dependent on the promotion of different mitigation and/or compensation measures, which can range from local layout adjustments, restrictions in construction scheduling or promoting regional habitat management actions. Ultimately, if the predicted impacts are too severe, it may lead to the entire project being cancelled.

In order to assess the potential impacts of wind farms on wolves, Portuguese researchers have relied on an array of wolf survey designs and techniques, adapted to the mild climate and absence of snow cover of this southern European country, which we summarise below.

5.2.2.1 Sampling Designs

BACI (before-after-control-impact): survey designs have been the standard for wolf monitoring programmes on wind farms in Portugal. Ideally, these studies are conducted from at least one full year before construction, throughout the whole construction period and for 3–5 years of the wind farm operation. The main goal is to compare several wolf biological parameters between time periods (before, during and after construction) along different distance gradients. This zonation includes control areas (defined as extents far enough from the wind farm to not be affected by the project but located close enough to be used by the same animals or pack) to assess whether any changes are due to the project or related to wolf population hazards. In BACI analysis, the percentage of biological indicators that are significantly different (positive or negative) when tested at a given significance level, is used to determine the direction and magnitude of the impact. Study areas are normally centred on the wind farms and then divided into square cells of 2 $\times$ 2 km or 5 $\times$ 5 km where wolf surveys techniques are systematically conducted (Fig. 5.3).

Impact–gradient approach: has been also used in wolf monitoring programmes on wind farms. This study design is used for impact quantification in relatively small assessment areas on homogeneous environments. The analysis is based on the relationship between the impact indicator (such as the level of wolf presence or abundance) and a gradient of distances from the wind turbines (concentric distances) (Fig. 5.4).

5.2.2.2 Wolf Biological Parameters

Presence/absence: this is the basic parameter for evaluating whether wolves use an area. Within each sampling unit (5 $\times$ 5 km or 2 $\times$ 2 km cell), wolf presence is surveyed throughout the whole monitoring plan (before, during and after construction) and variations in this parameter can be correlated with the wind farm proximity, construction schedule or human disturbance.

Relative abundance and spatial use: the intensity of use of each sampling unit can tell us more than just the presence or absence of the species. Wolf monitoring plans rely on relative abundance indexes to acknowledge changes in the use of space by this carnivore in relation to the wind farm development. The relative abundance indexes most commonly applied are kilometric abundance indexes

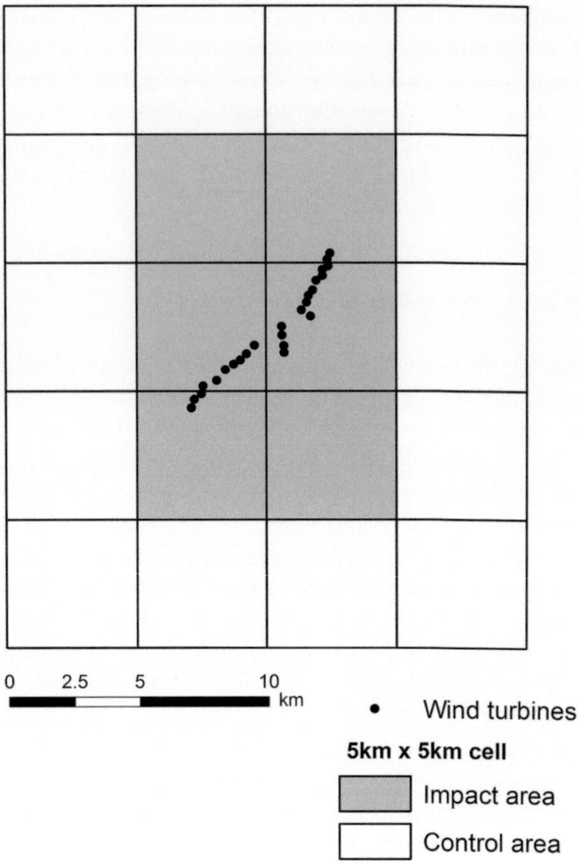

Fig. 5.3 Example of a BACI study area, centred on the wind farm and using 5 × 5 km cells. For posterior analysis, core cells around the wind turbines area considered impact areas while peripheral cells are considered control areas

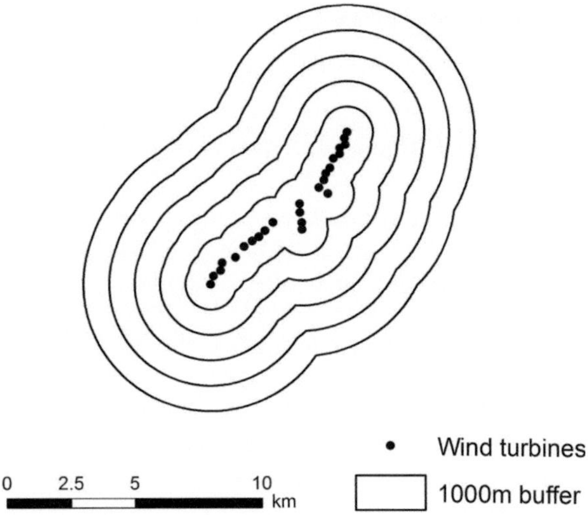

Fig. 5.4 Example of an impact–gradient study area, where concentric buffers of 1000 m width are used for analysis. Inner buffers are considered impact areas while marginal buffers are control zones

(KAI) for scat surveys or records abundance indexes (RAI) for photo trapping. The survey effort must be homogeneous along the whole study area so that results can be compared. Significant statistical differences between different cells or between seasons or years can be correlated with the presence of the wind farm (distance, construction schedule, operational phase).

Breeding location and success rate: wolves breed at specific locations where pups stay for a long period (3–5 months) in the dependency of adults. These sites are often reused over consecutive years and over several decades if no major disturbance event occurs (Mech and Boitani 2003). In this context, breeding sites, comprising both dens and pup-rearing areas, are considered crucial to assure wolf population viability in the short term. Differences in the location and use of these sites, as well as the reproductive success rate of the local pack can be correlated with the wind farm distance and operational phase (before, during and after construction).

5.2.2.3 Wolf Monitoring Techniques

Scats surveys: are the most common method for wolf monitoring in the Iberian Peninsula to assess wolf presence, relative abundance and intensive territorial marking as a predictor of wolf reproduction (e.g., Álvares et al. 2000; Roque et al. 2001; Eggermann et al. 2011; Blanco and Cortés 2012; Llaneza et al 2014). Therefore, this sampling method has been the most used for wolf monitoring programmes on wind farms in Portugal (Fig. 5.5a). Through predefined transects within each sampling unit, wolf scats are used as an indicator of the presence and abundance of this species. The Kilometric Abundance Index (KAI) is calculated monthly, seasonally and annually and then compared between time periods and between sampling units. Currently, most studies use genetic validation of detected scats, since the frequent presence of domestic dogs roaming within wolf range may lead to misidentifications on the field. The use of non-invasive genetic analysis also allows for the identification of individuals and determination of sex ratios, the minimum number of individuals in the study area and the kinship relations between animals, in order to better assess the local wolf population size and social structure.

Camera-trapping: has been used in recent years as a method to assess wolf presence and relative abundance in relation to wind farm projects (e.g., Grupo Lobo 2013). Using the same 2 × 2 km or 5 × 5 km sampling units, one or two cameras are deployed in each cell, and are on the field for 30–45 consecutive days within each survey period (Fig. 5.5b). The data are used to assess wolf presence at each sampling cell, and relative abundance indexes can be calculated using the number of independent events per day. An estimation of site occupancy can also be made using this methodology. Camera-trapping can also be deployed in certain areas to confirm breeding occurrence and determine breeding site location. Despite the advantages and usefulness of this method, there are also some drawbacks—namely the difficulty of using this method in poorly forested landscapes that are frequently used by people, leading to repeated camera thefts. On the other hand, is important to note that wolf photos or videos provided by camera-trapping have a strong

Fig. 5.5 Wolf sampling methods used in wind farm monitoring programmes in Portugal: **a** scats surveys (Photograph © 2008 Gonçalo Ferrão da Costa); **b** camera-trapping (Photograph © 2011 Gonçalo Ferrão da Costa); **c** live-capture for telemetry (Photograph © 2008 Monia Nakamura); **d** howling stations (Photograph © 2006 Gonçalo Ferrão da Costa)

communicational and awareness power close to stakeholders (such as wind farm promoters or Nature Conservation authorities), by providing visual evidences of wolf presence which often are crucial to support decision-making processes for mitigation measures, which other methodologies can hardly achieve (Fig. 5.6).

Telemetry is also being used locally to assess the responses of wolves to wind farms (Rio-Maior et al. 2010; Roque et al. 2011; Nakamura et al. 2013). This approach involves live-capturing wolves and fitting them with a GPS (Global Positioning System) collar in order to obtain fine-scale data on movements, activity and habitat use in relation to wind farm construction schedules (Fig. 5.5c). This method also has an important role to support decision-making processes for

Fig. 5.6 Iberian wolf recorded in camera-trapping during a wind farm monitoring plan, allowing an image with high potential for awareness (Photograph © 2013 Grupo Lobo)

mitigation measures, by showing to stakeholders fine-scale movements and behaviour of individual wolves.

Howling stations: are the most common method used to evaluate pack breeding success and to locate pup-rearing areas. Howling stations are conducted during the late breeding season, corresponding to the pup-rearing period (July–September) and use simulated howls to induce a response by the pack members (Fig. 5.5d). When a response is obtained, acoustic analysis of vocalisations makes it possible to discriminate between pups and adults, and, consequently, to confirm reproduction and breeding site location. Direct observations and camera-trapping can be conducted near known breeding areas, in order to detect pups and count a minimum number of individuals. These methodological approaches are normally applied where sign surveys, howling stations or telemetry reveal evidences of breeding sites. Changes in den locations and reproductive rates related to wind farms construction schedules are some of the most important effects to be studied during monitoring programmes and that theirs negative effects must be mitigated.

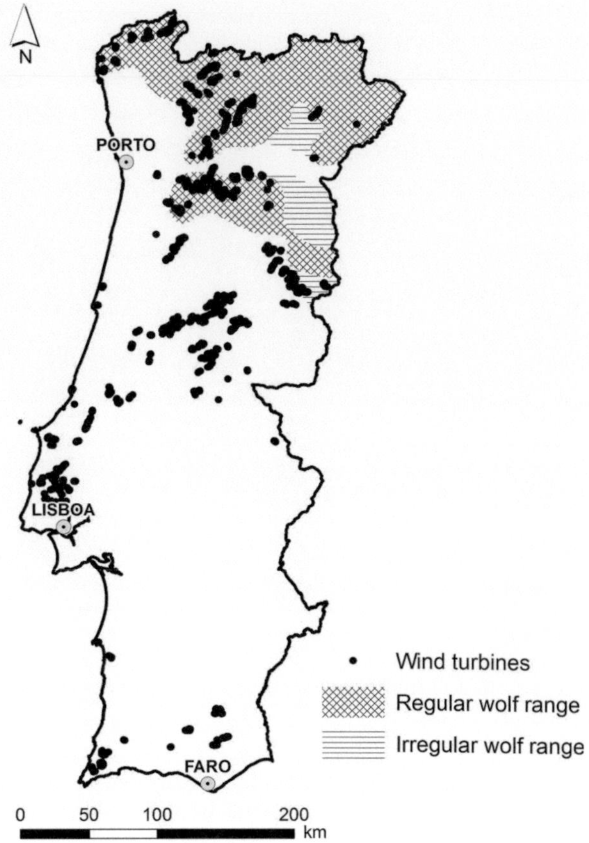

Fig. 5.7 Wolf range and wind turbines distribution in Portugal. Approximately 990 turbines were installed inside the Portuguese wolf range by the end of 2015

5.2.3 Effects of Wind Farms on Wolves in Portugal

Wind farm development in Portugal has an extensive overlap with the wolf distribution area. At the end of 2015, there were 990 wind turbines located inside the Portuguese wolf range (Fig. 5.7), covering the territories of 22 known wolf packs, around one third of the total number of packs estimated to occur in Portugal. These values correspond to an average of 4.8 wind turbines per 100 km^2 inside wolf range and may reach up to 120 wind turbines inside some pack territories located in the South Douro river subpopulation (Álvares et al. 2011) (Fig. 5.8) (for comparison, in Portugal, wolf density ranges from 0.5 to 3.0 wolves/100 km^2 and the average wolf territory size is 170 km^2; Pimenta et al. 2005; Álvares et al. 2015).

Due to the construction of these wind farms, several wolf monitoring programmes have been conducted in Portugal for the last 15 years. Despite the large amount of studies, most of the data collected and analysed is available only in technical reports written in Portuguese rather than in international scientific publications (but see Álvares et al. 2017).

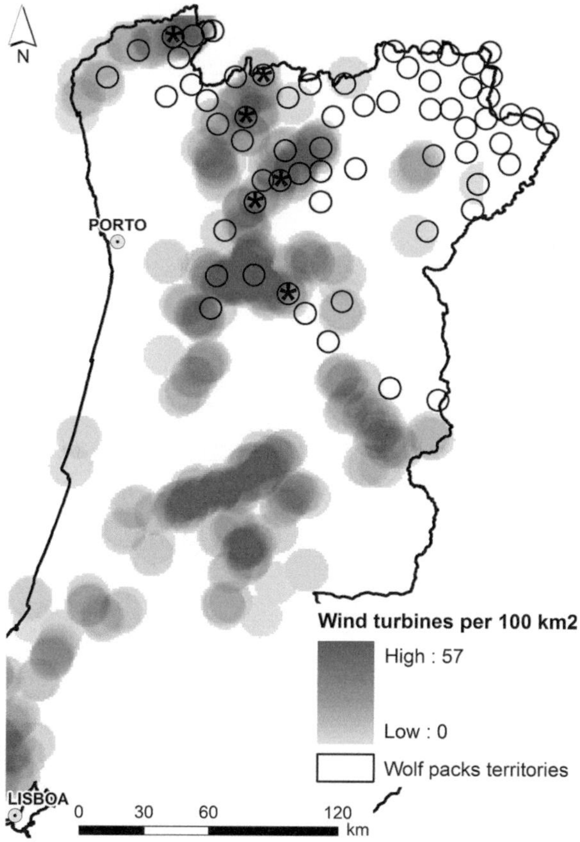

Fig. 5.8 Location of wolf packs in Portugal and wind turbines density per 100 km^2. In total, 22 known wolf territories have wind turbines, reaching 120 turbines inside some territories in the South Douro river subpopulation. Packs marked with * were used below as case studies to assess wolf breeding patterns (e.g., reproductive rates and breeding site location) in relation to wind farms

Based in the results of 11 wolf monitoring programmes conducted by the authors of this book chapter (Table 5.1, see Appendix), the effects of wind farms on wolves can be due to habitat disturbance during construction; acoustic and visual disturbance from wind turbines in operation and increasing circulation of vehicles in the road network built for wind-power development. Regarding disturbance from traffic, results from monitoring plans conducted in Portugal show an evident increase in the number of vehicles using the road network for wind farms. This is very evident during construction and the first years of operation, in contrast with the initial situation when those areas, normally located in remote mountain ridges, were mostly inaccessible to vehicles (Fig. 5.9 and 5.10). The recorded traffic was, on average, 36 times higher during the construction period than in pre-construction (and occasionally reached 200 times higher in some wind farm areas), reducing to 11 times higher in the third year of operation.

The recorded traffic of vehicles during construction and operation of wind farms frequently occurs during night and twilight periods, which largely coincides with the circadian period of highest wolf activity (Mech and Boitani 2003). The traffic inside wind farms is related not only to vehicles from construction and maintenance

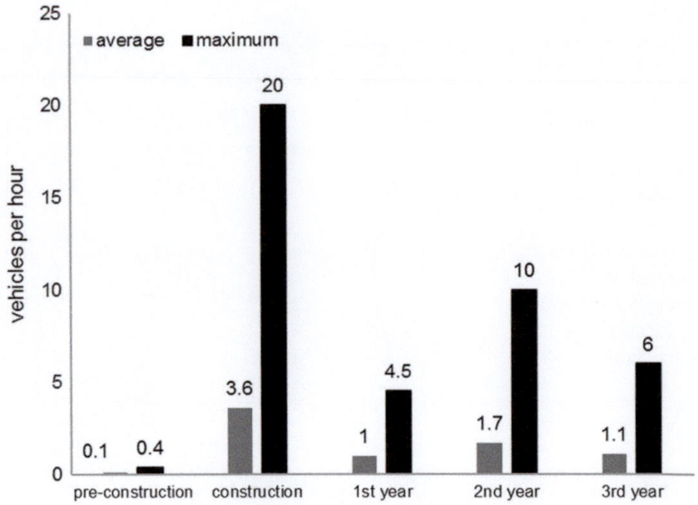

Fig. 5.9 Road traffic (measured as average and maximum number of vehicles per hour) recorded on wind farm areas according to wolf monitoring plans at 11 wind farms within wolf range: pre-construction (1 year), construction and operation (first 3 years)

Fig. 5.10 Examples of wind farm implementation areas within the Portuguese wolf range (Photographs © 2014, 2015, 2008, 2008 Gonçalo Ferrão da Costa)

of these infrastructures but also to vehicles from recreation, hunting, leisure and other outdoor activities. Traffic related to hunting activity is particularly frequent during the operation phase and often includes non-working days, such as weekends (Rio-Maior et al. 2010). Although utility traffic related to wind farm personnel is impossible to prevent, the circulation of leisure traffic is possible to minimise by restricting access for the general public (see Sect. 5.2.4). The augmentation of vehicles circulating in wind farm areas, due to access creation, promotes increased human disturbance, which may induce behavioural changes, particularly in elusive species such as wolves, and increases the risk of mortality due to car collision or poaching (Fig. 5.10a–d).

Regarding wolf presence, monitoring results have shown that this carnivore avoids wind farm areas during construction phase and, on some occasions, throughout the first years of operation. However, the effect seems to be limited as wolves continue to use areas where wind farms have been constructed. In fact, there are packs in Portugal that are still present, although with a very low reproductive rate, in territories where several different wind farms have been implemented (Álvares et al. 2011). The degree of use of wind farms areas by wolves appears to depend on the location and number of wind turbines, habitat suitability and proximity to important core areas in packs territories, such as breeding sites.

Results from wolf breeding patterns showed that packs already breeding more than 3 km away from wind farm areas during pre-construction periods had only minor changes in breeding site location and reproduction success (Fig. 5.11). However, when wind farms were built closer (<3 km) from breeding sites that were regularly used, wolves showed a decrease in breeding success during construction and initial operation phase, as well as shifts in denning sites, which were located progressively further from the wind farm (up to >6 km) to resume regular reproduction (e.g. Álvares et al. 2017) (Figs. 5.11, 5.12 and 5.13). This is particularly evident in the initial stage of wind farm operation, with less than 50% of the wolf packs studied being capable of reproduction in the first year (Fig. 5.13). Results also showed that after 3 years of operation, most of the packs were able of resume reproduction, but with shifts in denning location that involved, on average, a displacement of 2761 m from wind turbines, and could reach up to 6400 m in some packs (Figs. 5.11 and 5.12).

Such responses raise conservation concerns, particularly where availability of suitable breeding sites is a limiting factor and cumulative effects of other threats (e.g., additional infrastructure, human-related wolf mortality) may affect the local wolf population (Passoni et al. 2017). This is particularly relevant in endangered wolf populations occurring in highly humanised and heterogeneous landscapes, such as Portugal.

In conclusion, results from the monitoring studies conducted in Portugal show that wind farms induce important effects on wolves, such as: (i) changes in space use by avoidance during the construction and early operation phase; (ii) decreases in reproductive rates; and (iii) changes in the selection and fidelity of breeding sites used during the birth and pup-rearing period. However, on the other hand, there is evidence that newly formed packs, recently recolonising areas with already built wind farms, have shown a relative tolerance to these infrastructures, selecting breeding sites less than 3 km from wind turbines. Considering these

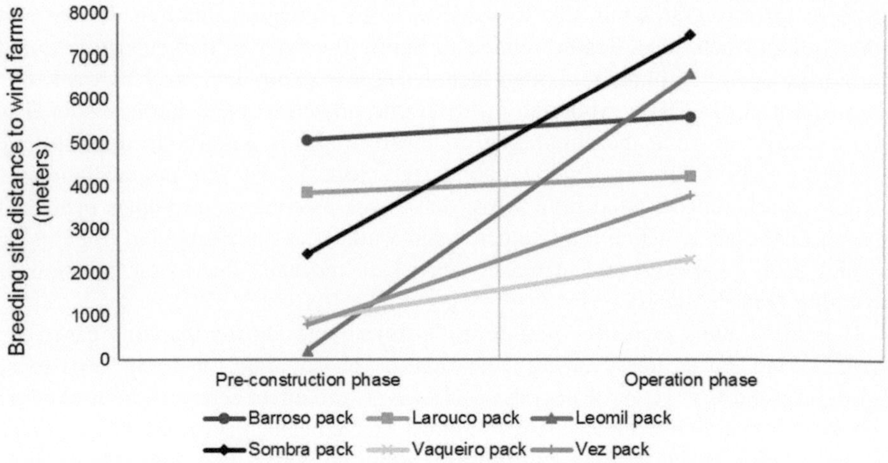

Fig. 5.11 Shifts in individual packs breeding locations, measured as average distance of the last 3 years before wind farm construction (pre-construction phase based in the projected area) and the first 3 years after wind farm construction (operation phase). Results were obtained from long-term monitoring of six wolf packs affected by wind farms comprising a total of 181 wind turbines

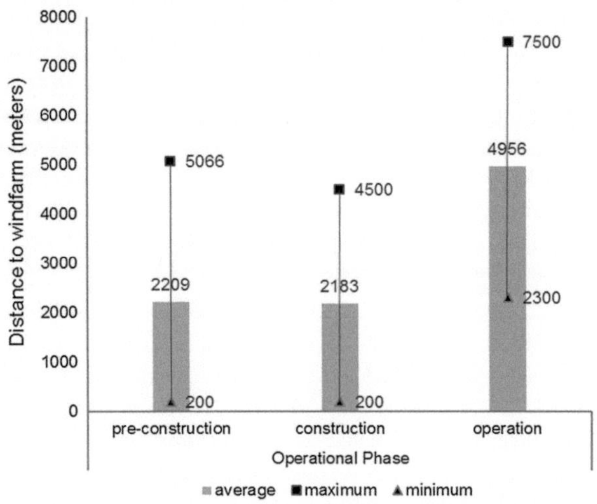

Fig. 5.12 Distances of wolf breeding sites to wind turbines (measured as average, maximum and minimum linear distances in meters) during wind farms schedule, including pre-construction, construction and operation phase, with shifts to breeding sites located further away from wind farms after the construction period. Note that distances reported for pre-construction phase are in relation to projected wind turbines. Results were obtained from long-term monitoring of six wolf packs affected by wind farms comprising a total of 181 wind turbines

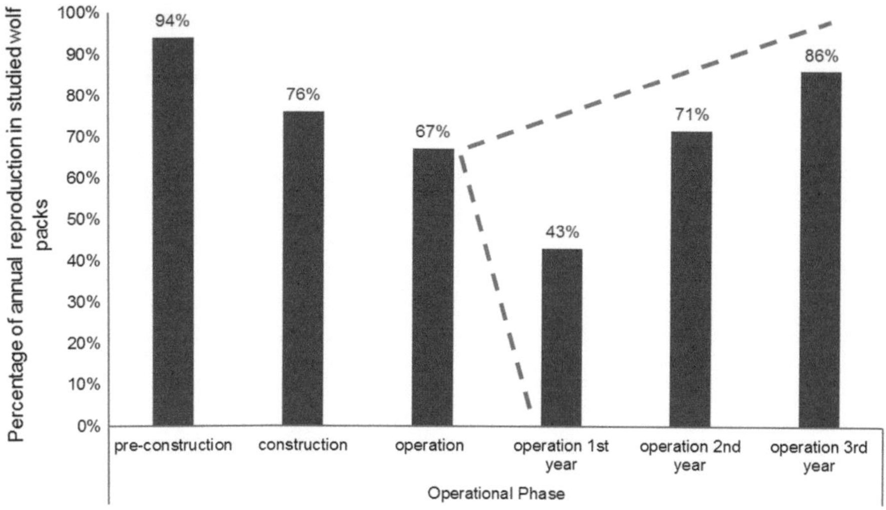

Fig. 5.13 Wolf reproduction success rates (measured as percentage of annual reproduction in studied packs) during wind farms schedule, including pre-construction, construction and operation phase, and in particular, during the first 3 years of operation, with an evident decrease from pre-construction period until early operation phase. Results were obtained from long-term monitoring of six wolf packs affected by wind farms comprising a total of 181 wind turbines

complex responses from wolves to wind farm development, a precautionary approach for protecting known wolf breeding sites should be adopted (see Sect. 5.2.4) in order to minimise the negative effects of wind farms on wolves. Further long-term monitoring studies should also be conducted to evaluate both the degree of tolerance of wolves to existing wind farms and the role of cumulative effects from other human activities on wolf population viability.

5.2.4 Lessons from the Portuguese Experience

The legal protection of wolves in Portugal, which makes them focal species in EIA studies of infrastructures, together with the recent development of wind energy in this country leads to a large number of wind farm monitoring plans targeting this carnivore, probably with no parallel anywhere in the world. Over the last 15 years, survey designs, methodologies or mitigation measures have been fine-tuned so that effective results could be achieved in a way that makes the most sense from a cost–benefit perspective.

In terms of methodological approaches, several considerations can be made:

- Sign surveys, conducted systematically in the study area, are an efficient standard methodology to assess changes on wolf presence and spatial use in relation to wind farms, but require genetic validation of scats, especially in areas with a low wolf density and/or where free-ranging dogs are common;

– Camera-trapping is also a valid methodology to evaluate wolf presence, spatial use and breeding occurrence, but its efficiency is largely dependent on the risk of being stolen. For example, at mountain ridges where most of the wind farms are built in Portugal, the habitat is dominated by open areas and low scrublands, where cameras cannot be hidden or secured from human robbery. The high number of thefts led to data loss and high financial costs.
– Telemetry is the most reliable method to assess fine-scale changes in wolf movements and spatial use. However, the duration and number of individuals that are tagged should be considered in order to cover the whole wind farm schedule (pre-construction, construction and operation phases) and evaluate possible individual differences related to age and social status.

Results have shown that the main impact of wind farms on wolves is the induced reduction on breeding site fidelity and reproductive rates. These effects, particularly when breeding sites shift to more unsuitable areas, may imply decreasing survival and pack viability in the short term. Thus, the primary focus of a monitoring plan should be the detection of breeding sites and breeding success rates before, during and after the construction of a wind farm.

Based on these findings, several mitigation and compensation measures have been applied to attenuate wind farm impacts on wolves in Portugal, such as:

– Protection of known breeding sites by considering a buffer at least 2 km from breeding locations as a 'no-construction' area for wind farm elements, including wind turbines, cables, road network and powerlines.
– Closing the road network built for wind power development in order to reduce traffic and direct human disturbance. However, in Portugal, the use of gates in the main access to wind farms has been difficult to implement due to public use of those areas and the claim for free access to mountain ridges by local populations.
– Besides other mitigation measures to reduce significantly the impacts of a wind farm (by layout adjustments or work restrictions during wolf breeding season or circadian periods with higher wolf activity), compensation measures have been also implemented to positively promote wolf presence at the local or even regional scale. These measures are mostly related to habitat management in order to promote suitable conditions for wolf breeding sites and to recover wild prey populations, such as roe deer (Brotas et al. 2015).
– Other compensatory measures aim to decrease the level of conflict from humans towards wolves, which is mainly related to livestock depredations, and includes actions such as promoting damage prevention measures by using livestock guarding dogs or electric fences (e.g., Ribeiro and Petrucci-Fonseca 2016).

Wolves are an adaptable species, and their ability to coexist with humans for centuries in a landscape as human-dominated as Portugal has been the key factor allowing their persistence despite intense human persecution. However, wolves can cope with human activities and disturbance only to a certain level and in the context of an increasing anthropogenic interference on worldwide habitats, the recent wind farm development can become a major concern when other cumulative factors also occur. Further research on the effect of wind farms in wolves is still needed but the results from monitoring programmes conducted in Portugal already provide valuable insights on the concerns and recommendations that should be taken into account when developing wind farm projects within the wolf range.

Acknowledgements We would like to thanks to everyone whose work contributed to the reviewed data, namely Sara Roque, Helena Rio-Maior, Monia Nakamura, Pablo Sierra, Julia Eggermann, Ana Guerra, Mónica Almeida, Diana Barreto and Barbara Marti-Domken. For financial support, we would like to thanks to EDP Renováveis, ENEOP2, Vento Minho SA and ACHLI.

Appendix

Table 5.1 Wolf monitoring projects used to assess wind farm effects in Portugal, namely for quantification of road traffic and evaluation of breeding patterns

Wind farm	Wind turbines	Wolf pack affected	Reference
Alto do Marco	5	Sombra	Bio3 & Grupo Lobo (2014). Monitorização do lobo-ibérico nos Parques Eólicos do Alto do Marco e Meroicinha II [Iberian wolf monitoring at Alto do Marco and Meroicinha II wind farms]. Relatório 5, Fase 3. 69 pp. (In Portuguese)
Alto Minho I	49	Vez	Álvares, F., H. Rio-Maior, S. Roque, M. Nakamura & F. Petrucci-Fonseca (2017). Ecological response of breeding wolves to wind farms: insights from two case studies in Portugal. In M.R. Perrow (Ed.), Wildlife and Wind Farms: Conflicts and Solutions. Volume 1: Onshore. -Volume 1: Onshore: 432 (pp. 225–227). Pelagic Publishing
Cabeço Alto	9	Larouco	Álvares, F. (2011). Ecologia e Conservação do lobo, *Canis lupus* L., no Noroeste de Portugal [Ecology and Conservation of *Canis lupus*, L. in Northwestern Portugal]. PhD Dissertation. Faculty of Sciences, University of Lisbon. 193 pp. (In Portuguese)
Falperra	20	Falperra/Sombra	Grupo Lobo (2016). Plano de Monitorização do Lobo no Parque Eólico de Falperra-Rechãnzinha, Ano 2 da fase de exploração [Wolf monitoring plan at Falperra/Rechãnzinha wind farm, 2nd year of operation phase], 56 pp (In Portuguese)
Meroicinha II	6	Sombra	Bio3 & Grupo Lobo (2014). Monitorização do lobo-ibérico nos Parques Eólicos do Alto

(continued)

Table 5.1 (continued)

Wind farm	Wind turbines	Wolf pack affected	Reference
			do Marco e Meroicinha II [Iberian wolf monitoring at Alto do Marco and Meroicinha II wind farms]. Relatório 5, Fase 3. 69 pp. (In Portuguese)
Negrelo-Guilhado	11	Falperra/Sombra	Ferrão da Costa & Petrucci-Fonseca (2011). Plano de Monitorização do Lobo – Parque Eólico de Negrelo e Guilhado [Wolf monitoring plan at Negrelo/Guilhado wind farm]. Relatório Técnico Anual 2010, abril de 2011. 36 pp. (In Portuguese)
Outeiro	13	Vaqueiro	Ferrão da Costa, G. & Álvares, F. (2008). Plano de Monitorização do lobo-ibérico no âmbito da Ampliação do Parque Eólico de Pena Suar – Ano 2007. (3º Ano da Fase II) [Iberian wolf monitoring plan at Pena Suar wind farm enlargement]. Relatório Técnico Final e Análise Global dos Resultados. ENERNOVA-EDP/CIBIO-UP. 46 pp. (In Portuguese)
Serra da Nave	19	Leomil	Álvares, F., H. Rio-Maior, S. Roque, M. Nakamura & F. Petrucci-Fonseca (2017). Ecological response of breeding wolves to wind farms: insights from two case studies in Portugal. In M.R. Perrow (Ed.), Wildlife and Wind Farms: Conflicts and Solutions. Volume 1: Onshore. -Volume 1: Onshore: 432 (pp. 225–227). Pelagic Publishing.
Serra do Alvão	21	Sombra	Saraiva, T., Petrucci-Fonseca, F., Ferrão da Costa, G., Barreto, D.& Marques, L. (2011). Monitorização do Parque Eólico da Serra do Alvão e Linha de Transporte de Energia [Wolf Monitoring anual report at Serra do Alvão wind farm]. Relatório Anual de Monitorização do Lobo-ibérico. Ecosativa, Lda. São Teotónio, January 2011. 32 pp.[In Portuguese]
Serra do Barroso III	11	Barroso	Ferrão da Costa (2013). Plano de Monitorização do Lobo – Parque Eólico da Serra do Barroso III [Wolf monitoring plan – Serra do Barroso III wind farm]. Relatório Final de Projeto. março 2013. 60 pp. (In Portuguese) Ferrão da Costa (2016). Plano de Monitorização do Lobo – Parque Eólico da Serra do Barroso III – Reforço de Potência [Wolf monitoring plan – Serra do Barroso III

(continued)

Table 5.1 (continued)

Wind farm	Wind turbines	Wolf pack affected	Reference
			wind farm - Repowering]. Relatório Anual, Ano2. julho 2016. 70 pp. [In Portuguese]
Vila Cova	17	Vaqueiro	Grupo Lobo (2016). Plano de Monitorização do Lobo no Parque Eólico de Vila Cova [Wolf monitoring plan at Vila Cova wind farm]. Ano 4, outubro 2016, 49 pp (In Portuguese]

References

Agha, M., Lovich, J. E., Ennen, J. R., Augustine, B., Arundel, T. R., & Murphy, M. O. et al. (2015). Turbines and terrestrial vertebrates: Variation in tortoise survivorship between a wind energy facility and an adjacent undisturbed wildland area in the desert southwest (USA). *Environmental Management, 56*(2), 332–341. doi:10.1007/s00267-015-0498-9

Álvares, F., Barroso, I., Blanco, J.C., Cortés, Y., Costa, G., Llaneza, L., Moreira, L., Nascimento, J., Palacios, V., Petrucci-Fonseca, F., Pimenta, V., Roque, S. & Santos, E. (2005). Wolf status and conservation in the Iberian Peninsula. *International Congress Frontiers of Wolf Recovery*, Colorado Springs, Colorado, USA, October 1–4, 2005.

Álvares, F., Barroso, I., Costa, G., Espírito-Santo, C., Fonseca, C., & Godinho, R. et al. (2015). Situação de referência para o Plano de Ação para a Conservação do Lobo-ibérico em Portugal. [Reference situation for the Wolf Conservation Action Plan in Portugal]. ICNF/CIBIO-INBIO/CE3C/UA, Lisboa, 70 pp (in Portuguese).

Álvares, F., H. Rio-Maior, S. Roque, M. Nakamura & F. Petrucci-Fonseca (2017). Ecological response of breeding wolves to wind farms: Insights from two case studies in Portugal. In M. R. Perrow (Ed.), Wildlife and wind farms: Conflicts and solutions. Volume 1: Onshore. - Volume 1: Onshore: 432 (pp. 225–227). Pelagic Publishing.

Álvares, F., H. Rio-Maior, S. Roque, M. Nakamura, D. Cadete, S. Pinto & F. Petrucci-Fonseca (2011). Assessing ecological responses of wolves to wind power plants in Portugal: Methodological constraints and conservation implications. In *Abstracts of CWW2011—conference on wind energy and wildlife impacts*, Trondheim, Norway, May 2–5, 2011

Álvares, F., Petrucci-Fonseca, F. & Pereira, E. (2000). O lobo no Parque Internacional Gerês-Xurés—Situação populacional e perspectivas de conservação [The wolf at Gerês-Xurês International Park]. *Galemys, 12*, 223–240.

APREN. (2017). *Technical data*. http://www.apren.pt/en/. Accessed February 20, 2017.

Baerwald, E. F., & Barclay, R. M. R. (2009). Geographic variation in activity and fatality of migratory bats at wind energy facilities. *Journal of Mammalogy, 90*(6), 1341–1349.

Berger, J. (2007). Fear, human shields and the redistribution of prey and predators in protected areas. *Biology Letters, 3*, 620–623.

Blanco, J. C., & Cortés, Y. (2012). Surveying wolves without snow: A critical review of the methods used in Spain. *Hytrix, 23*, 35–48.

Boitani, L., & Ciucci, P. (2009). Wolf management across Europe: Species conservation without boundaries. In M. Musiani, L. Boitani, & P. C. Paquet (Eds.), *A new era for wolves and people: Wolf recovery, human attitudes, and policy* (pp. 15–39). Calgary, Canada: University of Calgary Press.

Brotas, G., Oliveira, J. M., & Loureiro, C. (2015). Wind farms and a compensatory programme for the Iberian Wolf in Portugal: Seven years of analysis. In Abstracts of CWW2015—Conference on Wind Energy and Wildlife Impacts. Berlin, Germany, 10–12 March 2015.

Cabral, M. J., Almeida, J., Almeida, P. R., Dellinger, T., Ferrand de Almeida, N., & Oliveira, M. E. (2006). Livro Vermelho dos Vertebrados de Portugal [Portuguese Vertebrate Red Data Book]. Instituto de Conservação da Natureza, Lisboa. 660 p (in Portuguese).

De Lucas, M., Janss, G. F. E., & Ferrer, M. (2005). A bird and small mammal BACI and IG design studies in a wind farm in Malpica (Spain). *Biodiversity and Conservation, 14,* 3289–3303.

Dennehy, E., Llaneza, L., & López-Bao, J.V. (2013). Wolf *Canis lupus* persistence in context to road density and the changing concepts on wolf habitat characteristics. In Abstracts of the XI SECEM (Spanish Mammalogy Society) Congress, Avilés, Asturias, Spain, December 5–8, 2013.

Drewitt, A. L., & Langston, R. H. W. (2006). Assessing the impacts of wind farms on birds. *Ibis, 148,* 29–42.

Eggermann, J., Ferrão da Costa, G., Guerra, A. M., Kirchner, W. H., & Petrucci-Fonseca, F. (2011). Presence of Iberian wolf (*Canis lupus signatus*) in relation to land cover, livestock and human influence in Portugal. *Mammalian Biology, 76,* 217–221.

Elfstrom, M., Swenson, J. E., & Ball, J. P. (2008). Selection of denning habitats by Scandinavian brown bears. *Wildlife Biology, 14,* 176–187.

Flagstad, O., & Tovmo, M. (2010). Jerven pa Uljabuouda – hva viser DNAanalysene? [The wolverine at Uljabuouda – what does the DANN analyses show]. Mini report no. 305, NINA, Trondheim, Norway (in Norwegian).

Fuller, T. K., Berg, W. E., Raddle, G. L., Lenarz, M. S., & Joselyn, G. B. (1992). A history and current estimate of wolf distribution and numbers in Minnesota. *Wildlife Society Bulletin, 20,* 42–55.

George, S. L., & Crooks, K. R. (2006). Recreation and large mammal activity in an urban nature reserve. *Biological Conservation, 133,* 107–117.

Godinho, R., Lopes, S., & Ferrand, N. (2007). Estudo da diversidade e estruturação genética das populações de lobo (*Canis lupus*) em Portugal [Diversity and genetic strucure of wolf populations in Portugal]. Relatório Final. CIBIO/UP. 56 p (in Portuguese).

Grilo, C., Moço, G., Cândido, A. T., Alexandre, A. S., & Petrucci-Fonseca, F. (2002). Challenges for the recovery of the Iberian wolf in the Douro River south region. *Revista Biologia, 20,* 121–133.

Grupo Lobo. (2013). Plano de Monitorização do lobo-ibérico, Parque Eólico do Alto da Coutada, Relatório Final de Projeto [Wolf monitoring program at Alto da Coutada Wind farm, Final Report]. 88 p (in Portuguese).

Helldin, J. O., Jung, J., Neumann, W., Olsson, M., Skarin, A., & Widemo, F. (2012). The impact of wind power on terrestrial mammals: A synthesis. Swedish Environmental Protection Agency Report 6510. Stockholm, Sweden, 51 p.

Houle, M., Fortin, D., Dussault, C., Courtois, R., & Ouellet, J. P. (2010). Cumulative effects of forestry on habitat use by gray wolf (*Canis lupus*) in the boreal forest. *Landscape Ecology, 25* (3), 419–433.

Jedrzejewski, W., Niedziałkowska, M., Myslajek, R. W., Nowak, S., & Jedrzejewska, B. (2005). Habitat selection by wolves *Canis lupus* in the uplands and mountains of southern Poland. *Acta Theriologica, 50,* 417–428.

Kunz, T. H., Arnett, E. B., Erickson, W. P., Hoar, A. R., Johnson, G. D., Larkin, R. P., et al. (2007). Ecological impacts of wind energy development on bats: Questions, research needs, and hypotheses. *Frontiers of Ecology and Environment, 5,* 315–324.

Kuvlesky, W. P., Brennan, L. A., Jr., Morrison, M. L., Boydston, K. K., Ballard, B. M., & Bryant, F. C. (2007). Wind energy development and wildlife conservation: Challenges and opportunities. *Journal of Wildlife Management, 71,* 2487–2498.

Llaneza, L., Gracía, E. J., & López-Bao, J. V. (2014). Intensity of territorial marking predicts wolf reproduction: Implications for wolf Monitoring. *PLoS ONE, 9*(3), e93015. doi:10.1371/journal.pone.0093015

Llaneza, L., López-Bao, J. V., & Sazatornil, V. (2012). Insights into wolf presence in human-dominated landscapes: The relative role of food availability, humans and landscape attributes. *Diversity and Distributions, 18,* 459–469. doi:10.1111/j.1472-4642.2011.00869.x

Lopucki, R., & Mróz, I. (2016). An assessment of non-volant terrestrial vertebrates response to wind farms—a study of small mammals. *Environmental Monitoring and Assessment, 188*(2), 1–9. doi:10.1007/s10661-016-5095-8

Lovich, J. E., & Ennen, J. R. (2013). Assessing the state of knowledge of utility-scale wind energy development and operation on non-volant terrestrial and marine wildlife. *Applied Energy, 103,* 52–60.

May, R., Landa, A., van Dijk, J., Linnell, J. D. C., & Andersen, R. (2006). Impact of infrastructure on habitat selection of wolverines (*Gulo gulo*). *Wildlife Biology, 12,* 285–295.

Mech, L. D., & Boitani, L. (Eds.) (2003). *Wolves; Behavior, ecology and conservation.* The University of Chicago Press, 448 p.

Mech, L. D., Fritts, S. H., Raddle, G. L., & Paul, W. J. (1988). Wolf distribution and road density in Minnesota. *Wildlife Society Bulletin, 16,* 85–87.

Nakamura, M., Rio-Maior, H., Godinho, R., & Álvares, F. (2013). Investigação Aplicada à Conservação do Lobo no Noroeste de Portugal [Applied investigation to wolf conservation in northwestern Portugal]. Relatório Técnico – Ano II, CIBIO-UP, 56 p (in Portuguese)

Northrup, J. M., & Wittemyer, G. (2013). Characterising the impacts of emerging energy development on wildlife, with an eye towards mitigation. *Ecology Letters, 16,* 112–125.

Passoni, G., Rowcliffe, J. M., Whiteman, A., Huber, D., Kusak, J. (2017). Framework for strategic wind farm site prioritization based on modelled wolf reproduction habitat in Croatia. *European Journal of Wildlife Research.* doi:10.1007/s10344-017-1092-7

Pearce-Higgins, J. W., Stephen, L., Langston, R. H. W., Bainbridge, I. P., & Bullman, R. (2009). The distribution of breeding birds around wind farms. *Journal of Applied Ecology, 46,* 1323–1331.

Person, D. K., & Russell, A. L. (2009). Reproduction and den site selection by wolves in a disturbed landscape. *Northwest Science, 83*(3), 211–224.

Petrucci-Fonseca, F. (1990). O Lobo (*Canis lupus signatus* Cabrera, 1907) em Portugal - Problemática da sua Conservação [The Wolf in Portugal and its Conservation Problematic]. *Ph.D. Dissertation.* Faculty of Sciences, University of Lisbon (in Portuguese).

Pimenta, V., Barroso, I., Álvares, F., Correia, J., da Costa, G. F., & Moreira, L. et al. (2005). Situação populacional do lobo em Portugal: resultados do censo nacional 2002/2003 [Wolf population status in Portugal: Results from national censos 200/2003]. Instituto da Conservação da Natureza/Grupo Lobo, Lisboa, 158 p (in Portuguese).

Rabin, L. A., Coss, R. G., & Owings, D. H. (2006). The effects of wind turbines on antipredator behavior in California ground squirrels (*Spermophilus beecheyi*). *Biological Conservation, 131,* 410–420.

Ribeiro, S., & Petrucci-Fonseca, F. (2016). The use of LGDs under different husbandry systems in Portugal: Results from areas of continuous wolf presence and from newly recolonized areas. In Abstracts of the IV Wolf Iberian Congress, Castelo Branco, Portugal, October 27–30, 2016.

Rio-Maior, H., Nakamura. M., & Álvares, F. (2010). Programa de Monitorização do Lobo na área do Parque Eólico do Alto Minho I. Relatório Técnico Final. [Wolf Monitoring Program at Alto Minho I Wind farm area, Final Report], Veranda/CIBIO, 107 pp (in Portuguese).

Roque, S., Álvares, F., & Petrucci-Fonseca, F. (2001). Utilización espacio-temporal y hábitos alimentarios de un grupo reproductor de lobos en el noroeste de Portugal. *Galemys, 13,* 179–198 (In Spanish)

Roque, S., Godinho, R., Cadete, D., Pinto, S., Bernardo, J., & Petrucci-Fonseca, F. et al. (2011). Plano de Monitorização do Lobo Ibérico nas áreas dos Projectos Eólicos das Serras de Montemuro, Freita, Arada e Leomil [Iberian wolf monitoring plan in wind farm areas of Montemuro, Freita, Arada and Leomil mountains]. Relatório Técnico Final. CIBIO-UP/Grupo Lobo, 190 p (in Portuguese).

Santos, M., Bastos, R., Travassos, P., Bessa, R., Repas, M., & Cabral, J. (2010). Predicting the trends of vertebrate species richness as a response to wind farms installation in mountain ecosystems of northwest Portugal. *Ecological Indicators, 10,* 192–205.

Santos, M., Vaz, C., Travassos, P., & Cabral, J. (2007). Simulating the impact of socio-economic trends on threatened Iberian wolf populations *Canis lupus signatus* in north-eastern Portugal. *Ecological Indicators, 7,* 649–664.

Sazatornil, V., Rodríguez, A., Klaczek, M., Ahmadi, M., Álvares, F., Arthur, S., et al. (2016). The role of human-related risk in breeding site selection by wolves. *Biological Conservation, 201,* 103–110. doi:10.1016/j.biocon.2016.06.022

Taylor, L. (2010). The influences of roads on wolf movement on the Scandinavian peninsula in summer. Master thesis, Hedmark University, Faculty of Wildlife Management and Forestry, Evenstad.

Theuerkauf, J., Jedrzejewski, W., Schmidt, K., & Gula, R. (2003). Spatiotemporal segregation of wolves from humans in the Bialowieza Forest (Poland). *Journal of Wildlife Management, 67* (4), 706–716.

Thiel, R. R. (1985). Relationship between road densities and wolf habitat suitability in Wisconsin. *American Midland Naturalist, 113,* 404–407.

Torres, R. T., Silva, N., Brotas, G., & Fonseca, C. (2015a). To eat or not to eat? The diet of the endangered Iberian wolf (*Canis lupus signatus*) in a human-dominated landscape in central Portugal. *PLoS ONE, 10,* e0129379. doi:10.1371/journal.pone.0129379

Torres, R., Miranda, J., Carvalho, J. & Fonseca, C. (2015b). Expansion and current status of roe deer (*Capreolus capreolus*) at the edge of its distribution in Portugal. In *Annales Zoologici Fennici* (Vol. 52).

Vingada, J., Fonseca, C., Cancela, J., Ferreira, J., & Eira, C. (2010). Ungulates and their management in Portugal. In M. Apollonio, R. Andersen, R. J Putman (Eds,), Cambridge: Cambridge University Press.

Vos, J. (2000). Food habits and livestock depredation of two Iberian wolf packs (*Canis lupus signatus*) in the north of Portugal. *Journal of Zoology London, 251,* 457–462.

Wallin, J. A. (1998). A movement study of black bears in the vicinity of a wind turbine project, Searsburg, Vermont. Rapport till Green Mountain Power Corporation, South Burlington, Vermont, USA.

Walter, W. D., Leslie, D. M., Jr., & Jenks, J. A. (2006). Response of Rocky Mountain elk (*Cervus elaphus*) to wind-power development. *The American Midland Naturalist, 156,* 363–375.

Zimmermann, B., Nelson, L., Wabakken, P., Sand, H., & Liberg, O. (2014). Behavioral responses of wolves to roads: Scale-dependent ambivalence. *Behavioral Ecology, 25*(6), 1353–1364.

Chapter 6
Comparing Field Methods Used to Determine Bird and Bat Fatalities

João Paula, Margarida Augusto, Tiago Neves, Regina Bispo, Paulo Cardoso and Miguel Mascarenhas

6.1 Introduction

One of the impacts of wind farms is the direct mortality of flying vertebrates due to collision with the turbines (see Pereira et al. Chap. 2) (Dai et al. 2015; Drewitt and Langston 2008). To determine the magnitude of this impact on wild populations, animal carcasses are searched beneath the turbines to assess the number of fatalities that occur during the operational phase of a wind farm. However, taking solely the number of observed carcasses may lead to real mortality being largely underestimated, since only a fraction of animals killed by turbines will fall into the defined search area, some carcasses may disappear between consecutive searches due to removal by scavengers or they may decay (decomposition before searches take place), and some carcasses may remain undetected (Bernardino et al. 2013, 2011;

J. Paula (✉) · M. Augusto · T. Neves · P. Cardoso · M. Mascarenhas
Bioinsight, Odivelas, Portugal
e-mail: joao.p@bioinsight.pt

M. Augusto
e-mail: margarida.a@bioinsight.pt

T. Neves
e-mail: tiago.n@bioinsight.pt

P. Cardoso
e-mail: paulo.c@bioinsight.pt

M. Mascarenhas
e-mail: miguel.m@bioinsight.pt

R. Bispo
Eco-Ethology Research Unit, ISPA, Lisbon, Portugal
e-mail: regina.bispo@ispa.pt

R. Bispo
Marine and Environmental Sciences Centre, University of Lisbon, Lisbon, Portugal

M. Mascarenhas et al. (eds.), *Biodiversity and Wind Farms in Portugal*,
https://doi.org/10.1007/978-3-319-60351-3_6

Erickson et al. 2005). Real mortality has to be estimated, considering the number of carcasses found, which has to be corrected by the probability of carcass persistence, and carcass detection—frequently mentioned as the correction factors (Sinclair and DeGeorge 2016; Smallwood 2013; Korner-Nievergelt 2011; Miller 2008; Johnson et al. 2003).

In this chapter, we intend to discuss and compare the field methodologies used in monitoring programmes to search for carcasses and assess the correction factors in order to estimate mortality resulting from the operation of wind farms (see Marques et al. Chap. 7).

It is well documented that poor evaluation of searching efficiency and carcass removal may lead to heavily biased mortality estimates (Huso and Dalthorp 2014a; Smallwood 2013; Korner-Nievergelt 2011; Johnson 2003; Morrison 2002). Reyes et al. (2016) found that the positive bias of mortality estimates increased with decreasing efficiency of searchers. On the other hand, Santos et al. (2017) found that a reduction in monitoring periods and a shortening in the frequency of searches could reduce bias in mortality estimates. However, it should be noted that inconsistency in evaluation methods makes it impractical to compare studies (Huso and Dalthorp 2014b; Zimmerling et al. 2013), obscuring the extraction of a more general and clear picture of the problem.

In fact, earlier studies took only a limited approach to mortality estimates, documenting collision numbers for a particular season and location and rarely considering approaches to deal with the factors influencing mortality estimates (Arnett et al. 2008; Erickson et al. 2005). Reports from Portuguese studies since the mid-2000 have shown the same inconsistency patterns in reporting mortality rates with limited or no reference to correction factors (ICNB 2010). However, in Portugal, like in other countries, studies concerning mortality in wind farms progressively adopted standard protocols (Rodrigues et al. 2015; Bernardino et al. 2013; Strickland et al. 2011; ICNB 2009; SNH 2009) for evaluation of carcass searches, carcass persistence and carcass detection, allowing for reliable comparisons of studies.

In Portugal, general guidelines for wind farm monitoring studies were compiled from accumulated experience in Environmental Impact Assessment processes, extensive expert debate and literature analysis (Rodrigues et al. 2015; APA 2010). To obtain unbiased mortality estimates in wind farms, the mortality monitoring programme has to define the methodologies in a search area around each turbine as well as a search method. These programmes also have to consider methodologies to assess carcass persistence and carcass detection considering the search method adopted.

The optimisation of methodologies resulting from the intense debate around this theme is discussed below, with a bibliographic review/comparison of the existing methodologies in order to provide support to decision-makers in the Portuguese context.

6.2 The Methodologies

6.2.1 Carcass Searches

6.2.1.1 Search Area

There are several perspectives about the optimal size of the search area beneath the turbines. Many guidelines are based on conjecture rather than empirical evidence. This fact reduces confidence in the collected data, and precludes meaningful comparisons between sites (Hull and Muir 2010; Arnett et al. 2008; Bach and Rahmel 2004; Durr and Bach 2004). Table 6.1 summarises some distances that are used for search areas around wind farm turbines.

According to some international guidelines, search areas are normally centred around the wind turbine to be searched. In some states of Canada, it is recommended that the search area should be 50 m, or at least half the maximum turbine height for both birds and bats. In the state of New York in USA, the search area for bird and bat carcass searches could be no less than 1.5 times the rotor diameter. South Africa has separate guidelines for birds and bats, where the search area should be defined based on a certain percentage of the tower height considering the distance from the blade's tip to the ground—75% forbirds and 50% for bats. For bats, in Europe, EUROBATS (Rodrigues et al. 2015) recommend a radius equal to the total height of the wind turbine or at least 50 m as search area for this group. In Spain, for both birds and bats, a search area with a radius of at least 10% more than the rotor diameter is recommended.

The Portuguese guidelines for impact assessments in wind farms (APA 2010) state that the radius of the search area must be adjusted according to the wind turbine size (tower height and blade diameter), recommending a radius equal to a tower height of 5 m more than blade length. In ICNB (2009), which is more specific to bats, the recommended distance for bat carcass searches around the wind turbines is a minimum radius of 50 m, but ideally within a radius equal to the height of the wind turbine. According to the data obtained in Portugal, the majority of carcasses were detected within a maximum distance of 47 m from the turbine (ICNB 2010). However, it should be noticed that majority of this data is from mortality monitoring programmes implemented in Portugal where searching areas are set to 50 m around the wind turbine. This is a consequence of the wind turbines that are dominant in Portugal, with blade lengths less than 50 m (Simões 2015; Dias 2014; Lopes 2009).

EUROBATS convention supports that, ideally, the total area around the wind turbine should be searched, within a radius equal to the total height of the wind turbine, capturing available sample of birds and bats that are blown away by strong winds when collide (Rodrigues et al. 2015). Still, it is also known that vegetation height, among other obstacles, prevents a clean search around the area. In this situation, only the visible areas should be searched and, for each wind turbine,

Table 6.1 Summary of some guidelines of carcass searches worldwide

Country/Region (state if applicable)	Group	Search area	Search method	Frequency	Reference
Canada (Ontario and New Brunswick)	Birds and bats	50 m or at least half the maximum turbine height	Transects that are less than or equal to (<) 10 m apart	3–7 days depending on the season	Fish and Wildlife (2011), Ministry of Natural Resources of Ontario (2011)
USA (New York State)	Birds and bats	No less than 1.5 times the rotor diameter	Transects that are less than or equal to (<) 5 m apart	Part of the wind farm searched daily and the other part searched weekly	New York State Department of Environmental Conservation Division of Fish and Wildlife (2016)
South Africa	Birds	75% of the maximum distance from the blade's tip to the ground	Transects that are less than or equal to (<) 10 m apart	Search interval of 2 weeks	Jenkins et al. (2015)
	Bats	Half the maximum distance from the blade's tip to the ground	Transects that are less than or equal to (<) 6 m apart	Search interval of 1 week	Aronson et al. (2014)
Europe (all countries)	Bats	Radius equal to the total height of the wind turbine (at least 50 m)	Transects that are less than or equal to (<) 5 m apart	Search interval of 3 days	Rodrigues et al. (2015)
Spain	Birds and bats	At least 10% more than the rotor diameter	Transects that are less than or equal to (<) 5 m apart	Search interval of 2 weeks	Atienza et al. (2014)

(continued)

Table 6.1 (continued)

Country/Region (state if applicable)	Group	Search area	Search method	Frequency	Reference
Portugal	Birds and bats	Radius equal tower height or 5 m more than blade length	Transects that are less than or equal to (<) 5 m apart	Search interval of 7 days during wintering, nesting, juvenile dispersal and autumn migration and in the remaining months searches are on a monthly basis	APA, (2010)
	Bats	50 m radius or equal to the total height of the wind turbine.	Transects that are less than or equal to (<) 5 m apart	Search interval of 1 week from March to October	ICNB, (2009)

the percentage of the searched area must be calculated in order to correct the observed mortality (Rodrigues et al. 2015).

Ideally, all wind turbines on a wind farm should be searched for carcass detection, but large wind farms and/or financial and logistics limitations can lead to a random selection of wind turbines or a targeted sub-sample (Rodrigues et al. 2015) being searched. Strickland et al. (2011) recommended that in the USA at least 30% of wind turbines must be searched, with a minimum of 10 wind turbines. In South Africa, it is recommended that all turbines are surveyed.

In Portugal, the recommendations date from 2009 for bats and 2010 for birds and based on the existing European guidelines at that time. So according to (ICNB 2009) for bats, and (APA 2010) for birds, at least 70% of wind turbines should be surveyed. For bats, (ICNB 2009) followed the European guidelines at that time (Rodrigues et al. 2008), which proposed that the selection of wind turbines to monitor must include those near relevant habitats to bats like forests or rocky cliffs. However, according current European recommendations for bat carcass searches, every wind turbine in the wind farm should be sampled during each survey visit, and in the case of larger wind farms a subsample of the turbines may be randomly selected, stratified by habitat and/or wind farm features.

6.2.1.2 Search Method

Searches for carcasses around wind turbines have mostly been performed by human observers but recently the use of trained detection dogs to perform this task has emerged as an alternative methodology, since they have higher detection rates when compared with humans (Reyes et al. 2016; Beckmann et al. 2015; DeMatteo et al. 2014; Mathews et al. 2013; Paula et al. 2011; Reed et al. 2011, Arnett 2006, Homan et al. 2001, Smith et al. 2001).

Worldwide, transects see to be one of the most common methods recommended in national guidelines to preform carcass searches (Table 6.1). In South Africa, the guidelines for birds (Jenkins et al. 2015), recommend that the area around each turbine should be searched using linear transects no more than 10 m apart and 6 m apart for bats (Aronson et al. 2014), and that the search time be dependent on the size of the area and terrain conditions. In Europe, when carcass searches were performed by humans, it is recommended that the searches should be done in transects, separated 5 m apart (Rodrigues et al. 2015; Atienza et al. 2014; APA 2010; ICNB 2009). However, in difficult terrains with inaccessible zones, this requirement can be hard to fulfil (Santos et al. 2017). In these cases, search performed in quadrants, spiral or zig-zag transects or random walking is usually employed, allowing to the observer to adjust the search to the field conditions (Santos et al. 2017; Travassos et al. 2005).

In the Portuguese case, and according to guidelines for impact assessment in wind farms (APA 2010), a minimum of 20 min search is usually recommended depending of the dimension of the area to be searched and the vegetation type. According technical reports of monitoring programmes implemented in Portugal the zig-zag search is widely used. In one study carried out in Portugal, three methodologies were analysed: linear transects, zig-zag and capture–recapture and the results highlighted that the zig-zag methodology may be the most cost-effective (Soares and Bernardino 2014). On the other hand, in another recent study carried out in Portugal (Santos et al. 2017), results suggest that random walk searches are a viable methodology since when compared with spiral searches and parallel line searches methodologies, no significant differences in detection success were found. These authors also argue that random walk searches seem more adaptable to the real situations, specifically where marking track lines is unfeasible (or costly) and the structure of the vegetation or local orography prevent full spiral searches or parallel line searches.

The use of dogs to search for carcasses in wind farms is recent. When comparing both methodologies—carcass searches performed by humans or by dogs—it is evident that the dogs are more accurate and more efficient than humans (Reyes et al. 2016; Mathews et al. 2013; Paula et al. 2011). Results obtained by Reyes et al. (2016) lends support for using detection dogs to find rare fatalities when survey conditions give them an advantage over humans since they have a greater probability of detection, which increases the likelihood of finding these fatalities. However, the costs inherent to the dog team, which include training and maintaining the dogs, are higher. It must also be noted that the level of training is lower in humans, since the task is not complex and therefore it is relatively easy to form a person in order to perform this task. On the other hand, it is more complex to train a dog to detect bird and bat carcasses and to perform searches around wind turbines. Additionally, it should be considered the process of a handler formation that is also more complex than form a person to perform searches. As a recent methodology, there are no guidelines worldwide that recommend the best practices for applying this methodology. In Portuguese wind farms, Paula et al. (2011) used a methodology that followed the Portuguese recommendations for human searches. So, for a 50 m area they spent 10 min searching since dogs are more efficient than humans and can detect more carcasses in less time (Paula et al. 2011).

Independent of the search method adopted, the probability of encountering a carcass is dependent on the carcass detection by the technician or dog and the probability of carcass removal, which must be assessed by the different methodologies discussed in this chapter.

6.2.1.3 Search Frequency and Periods

The duration and frequency of carcass searches is greatly dependent on the monitoring objectives but is the same for both carcass search methodologies described in this section (humans and dogs). In Table 6.1, the search frequency recommended in some countries worldwide is summarised.

Duffy and Steward (2008) argue that a higher frequency of searching may produce more evidence of fatalities. However, the cost of this work can be of great concern since very intensive schemes can have a great monetary and logistical cost.

Therefore, to obtain a cost–benefit compromise, the scheme can either include or not all the seasons of the year and the search interval can be adjusted. In general, searches are focused in the most critical periods of mortality, such as, for example, in studies regarding bat mortality in which the searches are directed to the months when they are more active—between March and October—and so when the mortality rates are higher (ICNB 2009; Rodrigues et al. 2008; Kerns and Kerlinger 2004). In some states of Canada, the recommended frequency varies from 3–7 days depending on the season, and in the USA, more precisely in the state of New York, it is recommended that one part of the wind farm is searched daily and the other part searched weekly. In South Africa, a search interval of 2 weeks is recommended for birds (Jenkins et al. 2015), and 1 week for bats (Aronson et al. 2014), although this may change with each work objectives, field conditions and the carcass removal rates throughout the study. In Europe, the search schemes also vary with the available resources and monitoring objectives. For bats, according recent recommendations from EUROBATS (Rodrigues et al. 2015), carcass search must be performed every 3 days and should start as soon as bats become active after hibernation and last until they get back into hibernation, which should be adapted accordingly geographical and meteorological conditions of each region.

In the Portuguese case, the searches generally cover all months of the year, but with different frequencies such as weekly and monthly searches. According to the guidelines suggested by national authorities (ICNB 2009), the bat searches must be weekly between March and October (between November and February, is the hibernation period). Bat searches whenever possible must be carried along with bird searches in order to optimise human and logistic resources. The bird searches between November and February have to be carried out at least monthly (APA 2010). Recent investigation carried out in Portugal propose a reduction in the monitoring periods and daily searches in order to reduce bias in the estimations (Santos et al. 2017).

6.2.2 Carcass Removal Trials

To account for carcass removal, the observed mortality must be corrected by the probability of permanence of a carcass. To estimate this probability, wind farm monitoring plans include scavenger removal trials. Typically, in these trials, a certain number of carcasses are randomly placed underneath the wind turbines for an a priori fixed period of time. For each placed carcass, time until removal, i.e., time until a carcass is no longer available for detection (carcasses absent from the location of placement), is recorded. Most current and complete removal/decomposition trials consider different types of carcasses. Three representative size classes are frequently used associated with small, medium and large birds, and mice carcasses are commonly used as bats surrogates. Usually, carcasses are disposed of randomly. Sample size should be sufficient to enable statistical validation of the results. However, the majority of the studies are concerned with sacrificing the lowest number of animals or they use animals that died in wind farms. Another concern that is expressed in the majority of the studies is that the area should not be saturated with carcasses. This saturation can result as an artificial attractive for predators and could bias the results by skewing the normal speed at which the carcasses would be removed (Smallwood 2007; Travassos et al. 2005). For example, Smallwood et al. (2010) shows, in a study performed in a USA wind farm, that the speed of removal of the same carcass size is considerably different when compared with a higher and lower density situation in the placement of carcasses. Some studies reinforce the fact that carcass removal trials results are always site-specific and the permanence probability estimated for a specific wind farm should never be used to correct the observed mortality at another site (Bispo et al. 2013; Arnett et al. 2008). Therefore, experimental design must take into account factors that influence the removal rate, such as the season, carcass size and weather conditions (Paula et al. 2015; Bispo et al. 2013; Silva et al. 2008; Morrison 2002). In Portugal, it is recommended that carcass removal trials should be accomplished in each season (APA 2010; ICNB 2009).

Relative to the period that the carcasses were monitored for removal, studies commonly used four protocols: (i) daily checks for 7 days; (ii) daily checks for 14 days; (iii) daily checks for 20 days; (iv) daily visits in the first days after carcass placement, followed by visits spaced in time (Bispo et al. 2015; Bernardino et al. 2011). In Portugal, authorities recommend a daily check during 15 days (APA 2010; ICNB 2009). These recommendations were based on preliminary studies concerning carcass removal, which were carried out in three wind farms in Portugal, where results suggested a daily carcass check for a minimum period of 15 days in mountain ridges similar to those where the studied wind farms were located (Bernardino et al. 2011). In this methodology, the persistence time of each carcass is only known to lie within the time interval between two sequential inspections; such observations are said to be interval-censored (Bispo et al. 2013). In general, most of these trials are performed by human observers, who visit each carcass along the entire period or until it disappears. As such, this methodology involves significant human effort since daily trips must be made to the study site to check the carcasses until their removal.

To overcome some of the shortcomings with this methodology, such as costs and the fact that it provides only censored data, some recent studies point to the use of camera-trapping to study the removal of carcasses (Paula et al. 2015; Jones 2011; Smallwood et al. 2010; DeVault et al. 2004; Cunningham et al. 2002; Kostecke et al. 2001). In terms of human effort, this is a less demanding method since less frequent visits to the carcasses are required (Paula et al. 2015; Smallwood et al. 2010). Additionally, it allows the precise moment when a carcass was removed to be captured and the identification of the agent responsible for the removal (Paula et al. 2015). These results are leading the authorities to adopt this methodology in their recommendations. For example, in Scotland, the use of camera-trapping in removal trials is recommended in their official guide on methods for monitoring bird populations at onshore wind farms (SNH 2009). However, it should be noted that Bispo et al. (2015) found that conventional carefully planned removal trials that obtained censured data, in terms of total length and visits distribution over time, allowed estimates to be obtained that were as accurate as those obtained from camera-trapping trials where exact moment was recorded. According to Paula et al. (2015), some other issues should be taken into account when using this methodology since the higher initial investment in equipment means that it requires several trials for the outlay to be recovered. These authors also refer to the fact that equipment failure and time required for data analysis are important aspects to be taken into consideration by stakeholders when choosing camera-trapping techniques over the conventional method.

In Portugal, ICNB (2009) recommends the use of 20 carcasses for bat carcass removal trials. Paula et al. (2015) run the trials in two wind farms and use a total of 60 carcasses per trial, 30 birds (two size classes) and 30 mice (used as bats surrogates) to determine carcass removal. The total number of carcasses used by Bispo et al. (2013) in each trial varied between 20 and 80, according to the size of the 10 wind farms where the trials were carried out. Bernadino et al. (2011) used 10 carcasses of the three representative size classes associated with small, medium and large birds in each trial. Silva et al. (2008) used a total of 413 carcasses of four different types placed in three different lithological areas, with three different types of visibility (high, medium and low). The number of carcasses seems to be defined so that statistically robust results can be obtained.

6.2.3 Detection Trials

In the case of human observers, their detection accuracy and efficiency can be conditioned by numerous factors; for example, carcass size, carcass state of decomposition, vegetation cover, vegetation density, orography, weather conditions, time spent in each wind turbine and, in the latter case, by observer fatigue. On the other hand, as dogs rely exclusively on their olfactory skills to detect carcasses, their accuracy, if trained accordingly the best practices, is independent of carcass size, colour, state of decomposition, visibility due to vegetation height and density (Paula et al. 2011). However, their efficiency could be influenced by orography, weather conditions, and also, like humans, fatigue, which could affect accuracy (Paula et al. 2011). Therefore, detection trials should be made for each wind farm and, according to Portuguese authorities, should be performed at least once in each season (APA 2010).

For human searchers an experimental design that allows the integration of different observers and the different degrees of searched area is needed in order to obtain carcass detection probability. It is crucial to avoid an unnecessary number of sacrificed animals, so objects or models similar to small, medium and large carcasses have been used. Strickland et al. (2011) recommends that at least 50 carcasses/models for each combination of factors (carcass/model size, type of vegetation, etc.) should be used and, for Portugal, national authorities recommend at least 20 carcasses/models (ICNB 2009). During the trials, all areas are searched by observers using the same methods as for carcass searches. At the end, the number of carcasses found is recorded in order to determine the empirical detection rate (%) of the observer (e.g., Erickson et al. 2000). Estimating the empirical detection rate was simplified by implementing calculation procedures in the web-based application developed by Bispo et al. (2010), available at Wildlife Fatality Estimator platform (www.wildlifefatalityestimator.com).

In Portugal, the use of objects or models in detection trials is a normal procedure and is recommended by national authorities (APA 2010). Normally, three size classes are used. In the experimental design, two approaches have typically been used: one does not consider vegetation cover, and the other does consider this parameter. In areas where the vegetation is homogeneous, the vegetation cover is not considered as a variable and so models are randomly distributed around wind turbines and their surroundings. In this case, carcass size is the only variable taken into consideration. With this approach, the detection probability is calculated for the area where the wind farm is installed. On the other hand, in areas where vegetation is heterogeneous, this variable is taken into account. And so, the number of models used has to be representative of each size class and vegetation unit defined. With this approach, the detection probability of the area is obtained based on the representability of the vegetation. The number of objects or models used in monitoring programmes that are implemented in Portugal varies depending on the size classes considered and the homogeneity or heterogeneity of vegetation cover.

For dogs, according to the results of Paula et al. (2011), the trials do not need to consider vegetation cover and structure, carcass size or decomposition. Based on findings from Mathews et al. (2013), Paula et al. (2011), the design of detection trials for dogs should consider the use of bird and bat carcasses, if possible from wind farms or other infrastructures that cause mortality, placed along the wind farm in order to estimate the detection probability for the area. Since it has to be statistically validated, the number of carcasses recommended by the authorities is used, in this case considering type of carcass (bird or bat).

6.3 Conclusions

In Portugal, protocols for carcass searches, carcass removal and efficiency trials implemented suffer an evolution along the years especially since recommendations are published by national authorities, namely APA (2010), ICNB (2009). In fact, bat guidelines (ICNB 2009) are being revised according to European guidelines proposed by Rodrigues et al. (2015).

To Portugal, a search area that covers a minimum of 5 m more than the blade length seems to be effective to collect sufficient data to estimate mortality. To search the area for carcasses, in Portugal, studies seem to indicate that random walking or zig-zag methods are a cost-effective way to collect carcasses in operating wind farms. If there are no cost constraints, the use of dogs could be an alternative method in sensitive areas, where rare species susceptible to impacts are present. According to recent investigations (Santos et al. 2017) and recommendations

(Rodrigues et al. 2015), a frequency of one to three days seems to reduce bias in the estimations.

To perform removal trials, in Portugal, camera-trapping seems to be a cost-effective method to collect data if several trials have to be made in order to recover initial investment. This method also has the advantage of identifying the removal agent, contributing to the knowledge of the local communities where trials are carried out and, consequently, it contributes to mapping the species and so helps to evaluate impacts in other projects implemented in the area. However, if the initial investment does not compensate for the reduction of costs, the traditional method, if carefully planned, can also obtain estimates that are as accurate as the ones obtained from camera-trapping (Bispo et al. 2015).

For detection trials, in Portugal, the use of models with three sizes seems to be an effective way to obtain detection probability and a way to avoid the unnecessary sacrifice of animals. Detection trials should include visibility classes if vegetation is very heterogenous. If vegetation is homogeneous, models should be distributed along the area. The number of models that should be used varies depending on the size classes considered and the homogeneity or heterogeneity of vegetation cover, but must be sufficient to produce statistically robust results.

References

APA. (2010). Guia Metodológico para a Avaliação de Impacte Ambiental em Parques Eólicos. Agência Portuguesa do Ambiente.

Arnett, E. B. (2006). A preliminary evaluation on the use of dogs to recover bat fatalities at wind energy facilities. *Wildlife Society Bulletin, 34,* 1440–1445.

Arnett, E. B., Brown, W., Erickson, W. P., Fiedler, J. K., Hamilton, B. L., Henry, T. H., et al. (2008). Patterns of bat fatalities at wind energy facilities in North America. *The Journal of Wildlife Management, 72*(1), 61–78.

Aronson, J., Richardson, E. K., MacEwan, K., Jacobs, D., Marais, W., Aiken, S., et al. (2014). *South African good practice guidelines for operational monitoring for bats at wind energy facilities* (Vol., 1st ed.).

Atienza, J. C., Martín Fierro, I., Infante, O., & Valls, J. (2014). *Guidelines for assessing the impact of wind farms on birds and bats* (version 4.0). SEO/BirdLife, Madrid.

Bach, L., & Rahmel, U. (2004). Summary of wind turbine impacts on bats: assessment of a conflict. *Brem Beitr Naturk Naturs, 7,* 245–252.

Beckmann, J. P., Waits, L. P., Hurt, A., Whitelaw, A. & Bergen, S. (2015). Using detection dogs and Rspf models to assess habitat suitability for bears in greater yellowstone. *Western North American Naturalist, 75*(4), 396–405. doi:http://dx.doi.org/10.3398/064.075.0410

Bernardino, J., Bispo, R., Costa, H., & Mascarenhas, M. (2013). Estimating bird and bat fatality at wind farms: A practical overview of estimators, their assumptions and limitations. *New Zealand Journal of Zoology, 40*(1), 63–74.

Bernardino, J., Bispo, R., Torres, P., Rebelo, R., Mascarenhas, M., & Costa, H. (2011). Enhancing carcass removal trials at three wind energy facilities in Portugal. *Wildlife Biology in Practice, 7,* 1–14.

Bispo, R., Bernardino, J., Marques, T. A., & Pestana, D. (2013). Discrimination between parametric survival models for removal times of bird carcasses in scavenger removal trials at wind turbines sites. In J. Lita da Silva et al. (Eds.), *Advances in regression, survival analysis, extreme values, markov processes and other statistical applications, studies in theoretical and applied statistics*. Berlin, Heidelberg: Springer.

Bispo, R., Bernardino, J., Paula, J., Marques, T., Costa, H., & Mascarenhas, M. (2015). Enhancing the study design of carcass removal trials for bats and birds at wind farms. In *Conference on Wind energy and Wildlife impacts*, Berlin, Germany, March 10–12, 2015.

Bispo, R., Palminha, G., Bernardino, J., Marques, T., & Pestana, D. (2010). A new statistical method and a web-based application for the evaluation of the scavenging removal correction factor. In *Proceedings of the VIII Wind Wildlife Research Meeting* (pp. 33–35). Lakewood, Colorado, October 19–21, 2010.

Cunningham, P. D., Brown, L. J., & Harwood, A. J. (2002). *Predation and scavenging of salmon carcasses along spawing streams in the Scottish Highlands*. Final report for the Atlantic Salmon Trust.

Dai, K., Bergot, A., Liang, C., Xiang, W. N., & Huang, Z. (2015). Environmental issues associated with wind energy—A review. *Renewable Energy, 75,* 911–921.

DeMatteo, K. E., Rinas, M. A., Argüelles, C. F., Holman, B. E., Di Bitetti, M. S., Davenport, B., et al. (2014). Using detection dogs and genetic analyses of scat to expand knowledge and assist felid conservation in Misiones, Argentina. *Integrative Zoology, 9,* 623–639.

DeVault, T. L., Brisbin, I. L., Jr., & Rhodes, O. E., Jr. (2004). Factors influencing the acquisition of rodent carrion by vertebrate scavengers and decomposers. *Canadian Journal of Zoology, 82,* 502–509.

Dias, M. (2014). *Impacto Económico-financeiro no Sobreequipamento de Centrais Eólicas*. Tese de Mestrado em Engenharia Eletrotécnica e de Computadores: Faculdade de Engenharia da Universidade do Porto.

Drewitt, A., & Langston, R. H. (2008). Collision effects of wind-power generators and other obstacles on birds. *Annals of the New York Academy of Sciences, 1134,* 233–266.

Duffy, K. & Steward, M. (2008). *Turbine search methods and carcass removal trials at the Braes of Doune windfarm*. Natural Research Information Note 4. Natural Research Ltd., Banchory, UK.

Dürr, T., & Bach, L. (2004). Bat deaths and wind turbines: A review of current knowledge and of information available in the database for Germany. *Brem Beitr Naturk Naturs, 7,* 253–264.

Erickson, W., Johnson, G., Strickland, M. D., & Kronner, K. (2000). *Avian and bat mortality associated with The Vansycle wind project*. Umatilla County, Oregon. 1999 Study Year. Western EcoSystems Technology, Inc.

Erickson, W., Johnson, G., & Young, D. (2005). *A summary and comparison of bird mortality from anthropogenic causes with an emphasis on collisions*. Technical report, USDA Forest Service.

Fish and Wildlife. (2011). Post-construction bat and bird mortality survey guidelines for wind farm development in new Brunswick.

Homan, J. H., Linz, G., & Peer, B. D. (2001). Dogs increase recovery of passerine carcasses in dense vegetation. *Wildlife Society Bulletin, 29,* 292–296.

Hull, C. L., & Muir, S. (2010). Search areas for monitoring bird and bat carcasses at wind farms using a Monte-Carlo model. *Australasian Journal of Enviromental Managment, 17,* 77–87.

Huso, M. M. P., & Dalthorp, D. (2014a). Accounting for unsearched areas in estimating wind turbine-caused fatality. *The Journal of Wildlife Management, 78*(2), 347–358. doi:10.1002/jwmg.663

Huso, M. M. P., & Dalthorp, D. (2014b). A Comment on "Bats killed in large numbers at united states wind energy facilities". *BioScience*. doi:10.1093/biosci/biu056

ICNB. (2009). *Recomendações para Planos de Monitorização de Parques Eólicos—Quirópteros* (p. 10). Lisboa: Instituto da Conservação da Natureza e da Biodiversidade.

ICNB. (2010). Avaliação do efeito dos parques eólicos sobre os morcegos em Portugal continental (análise dos dados disponíveis em Outubro de 2009). Instituto da Conservação da Natureza e da Biodiversidade. Relatório não publicado.

Jenkins, A. R., van Rooyen, C. S., Smallie, J. J., Harrison, J. A., Diamond, M., Smit-Robinson, H. A., et al. (2015). *Best-practice guidelines for assessing and monitoring the impact of wind-energy facilities on birds in Southern Africa* (Vol., 3rd ed.).

Johnson, G. D., Erickson, W. P., Strickland, M. D., Sheperd, M. F., Sheperd, D. A., & Sarappo, S. A. (2003). Mortality of bats at a large-scale wind power development at Buffalo Ridge, Minnesota. *The American Midland. Naturalist, 150*(2), 332–342. doi:10.1674/0003-0031

Jones, A. (2011). *Animal scavengers as agents of decomposition: The post-mortem succession of Louisiana wildlife*. Master thesis, Louisiana State University.

Kerns, J., & Kerlinger, P. (2004). A study of bird and bat collisions at the mountaineer wind energy facility, Tucker County, West Virginia: Annual Report for 2003. Prepared for FPL Energy and the Mountaineer Wind Energy Center Technical Review Committee. February 14, 2004. Technical report prepared by Curry and Kerlinger, LLC., for FPL Energy and Mountaineer Wind Energy Center Technical Review Committee. Curry and Kerlinger, LLC (39 pp).

Korner-Nievergelt, F., Korner-Nievergelt, P., Behr, O., Niermann, I., Brinkmann, R., & Hellriegel, B. (2011). A new method to determine bird and bat fatality at wind energy turbines from carcass searches. *Wildlife Biology, 17*(4).

Kostecke, R. M., Linz, G. M., & Bleier, W. J. (2001). *Survival of avian carcasses and photographic evidence of predators and scavengers*. USDA National Wildlife Research Center —Staff Publications. Paper 510.

Lopes, A. (2009). *Produção Eólica e Enquadramento Técnico-Económico em Portugal*. Mestrado em Engenharia Electrotécnica e de Computadores. Faculdade de Engenharia da Universidade do Porto.

Mathews, F., Swindells, M., Goodhead, R., August, T., Hardman, P., Linton, D. M., et al. (2013). Effectiveness of search dogs compared with human observers in locating bat carcasses at wind-turbine sites: A blinded randomized trial. *Wildlife Society Bulletin, 37*(1), 34–40. doi:10.1002/wsb.256

Miller, A. (2008). *Patterns of avian and bat mortality at a utility scaled wind farm on the southern high plains*. Ph.D. thesis, Texas Tech University.

Ministry of Natural Resources of Ontario. (2011). *Birds and bird habitats: Guidelines for wind power projects*.

Morrison, M. (2002). *Searcher Bias and Scavenging Rates in Bird/Wind Energy Studies*. Technical Report NREL/SR-500–30876, National Renewable Energy Lab, Golden Colorado, 5.

New York State Department of Environmental Conservation Division of Fish and Wildlife. (2016). *Guidelines for conducting bird and bat studies at commercial wind energy projects*.

Paula, J. J., Bispo, R. M., Leite, A. H., Pereira, P. G., Costa, H. M., Fonseca, C. M., et al. (2015). Camera-trapping as a methodology to assess the persistence of wildlife carcasses resulting from collisions with human-made structures. *Wildlife Research, 41*(8), 717–725.

Paula, J., Leal, M. C., Silva, M. J., Mascarenhas, R., Costa, H., & Mascarenhas, M. (2011). Dogs as a tool to improve bird-strike mortality estimates at wind farms. *Journal for Nature Conservation, 19*, 202–208.

Reed, S. E., Bidlack, A. L., Hurt, A., & Getz, W. M. (2011). Detection distance and environmental factors in conservation detection dog surveys. *Journal of Wildlife Management, 75*(1), 243–251.

Reyes, G. A., Rodriguez, M. J., Lindke, K. T., Ayres, K. L., Halterman, M. D., Boroski, B. B., et al. (2016). Searcher efficiency and survey coverage affect precision of fatality estimates. *The Journal of Wildlife Management, 80*(8), 1488–1496.

Rodrigues, L., Bach, L., Dubourg-Savage, M.J., Goodwin, J., & Harbusch, C., (2008). *Guidelines for consideration of bats in wind farm projects*. EUROBATS Publication Series No. 3. UNEP/EUROBATS Secretariat, Bonn, Germany.

Rodrigues, L., Bach, M. J. Dubourg-Savage, B., Karapanza, D., Kovac, T., Kervyn, J., et al. (2015). *Guidelines for consideration of bats in wind farm projects—Revision 2014* (133 pp). EUROBATS Publication Series No. 6 (English version). UNEP/EUROBATS Secretariat, Bonn, Germany.

Santos, M., Bastos, R., Ferreira, D., Santos, A., Barros, P., Travassos, P., et al. (2017). A spatial explicit agent based model approach to evaluate the performance of different monitoring options for mortality estimates in the scope of onshore windfarm impact assessments. *Ecological Indicators, 73,* 254–263.

Silva, B., Barreiro, S., & Alves, P. (2008). Factors influencing carcass removal at wind-farms, in mountain areas of the center-north region of Portugal. In *Abstracts XIth European Bat Research Symposium* (pp. 18–22), Agosto, Roménia.

Simões, S. (2015). *Caracterização do Ruído produzido por um parque eólico. Efeito sobre a população.* Tese de Mestrado em Segurança e Higiene no Trabalho. Escola Superior de Ciências Empresariais—Escola Superior de Tecnologia de Setúbal.

Sinclair, K., & DeGeorge, E. (2016). *Framework for testing the effectiveness of bat and eagle impact-reduction strategies at wind energy projects.* National Renewable Energy Laboratory. Technical Report NREL/TP-5000-65624.

Smallwood, K. S. (2007). Estimating wind turbine-caused bird mortality. *The Journal of Wildlife Management, 71,* 2781–2791. doi:10.2193/2007-006

Smallwood, K. S. (2013). Comparing bird and bat fatality-rate estimates among North American wind-energy projects. *Wildlife Society Bulletin, 37*(1), 19–33. doi:10.1002/wsb.260

Smallwood, K. S., Bell, D. A., Snyder, S. A., & Didonato, J. E. (2010). Novel scavenger removal trials increase wind turbine—Caused avian fatality estimates. *The Journal of Wildlife Management, 74,* 1089–1096. doi:10.2193/2009-266

Smith, D. A., Ralls, K., Davenport, B., Adams, B., & Maldonado, J. E. (2001). Canine assistants for conservationists. *Science, 291,* 435.

SNH. (2009). *Guidance on methods for monitoring bird populations at onshore wind farms.* Guidance Note. Scottish Natural Heritage. Scotland, UK. Available from http://www.snh.gov.uk/

Soares, M., & Bernardino, J. (2014). T4.3.4 Aperfeiçoamento do desenho experimental dos testes de detectabilidade de cadáveres- Relatório final. Wind & Biodiversity—Projecto 11541.

Strickland, M. D., Arnett, E. B., Erickson, W. P., Johnson, D. H., Johnson, G. D., Morrison, M. L., et al. (2011). *Comprehensive guide to studying wind energy/wildlife interactions.* Prepared for the National Wind Coordinating Collaborative, Washington, D.C., USA

Travassos, P., Costa, H. M., Saraiva, T., Tomé, R., Armelin, M., Ramírez, F. I., et al. (2005). *A energia eólica e a conservação da avifauna em Portugal.* Lisboa: SPEA.

Zimmerling, J. R., Pomeroy, A. C., d'Entremont, M. V., & Francis, C. M. (2013). Canadian estimate of bird mortality due to collisions and direct habitat loss associated with wind turbine developments estimation de la mortalité aviaire canadienne attribuable aux collisions et aux pertes directes d'habitat associées à l'éolien. *Avian Conservation and Ecology, 8*(2), 10.

Chapter 7
Estimating Bird and Bat Fatality at Wind Farms: From Formula-Based Methods to Models to Assess Impact Significance

Joana Marques, Luísa Rodrigues, Maria João Silva, Joana Santos, Regina Bispo and Joana Bernardino

7.1 Introduction

Comprehension of the impacts caused by wind farms on local communities, namely on flying vertebrates communities, is the main objective of any monitoring programme. The best documented negative impacts on birds and bats in wind farms are collisions with wind turbines (e.g., Marques et al., Chap. 1, Pereira et al., Chap. 2).

J. Marques (✉) · J. Santos
Bioinsight, Odivelas, Portugal
e-mail: joana.m@bioinsight.pt

J. Santos
e-mail: joana.s@bioinsight.pt

L. Rodrigues
Instituto da Conservação da Natureza e das Florestas, Lisbon, Portugal
e-mail: luisa.rodrigues@icnf.pt

M.J. Silva
Centre for Environmental and Marine Studies, University of Aveiro, Aveiro, Portugal
e-mail: mjjds@st-andrews.ac.uk

M.J. Silva
School of Biology, University of St Andrews, St Andrews, UK

R. Bispo
Eco-Ethology Research Unit, ISPA, Lisbon, Portugal
e-mail: regina.bispo@ispa.pt

R. Bispo
Marine and Environmental Sciences Centre, University of Lisbon, Lisbon, Portugal

J. Bernardino
CIBIO/InBIO, Department of Biology, University of Évora, Évora, Portugal
e-mail: joanab@uevora.pt

M. Mascarenhas et al. (eds.), *Biodiversity and Wind Farms in Portugal*,
https://doi.org/10.1007/978-3-319-60351-3_7

Since the beginning of wind farms' operation, when this impact was detected in the USA (e.g., Orloff and Flanerry 1992), monitoring programmes started to include dedicated methodologies to assess and determine the magnitude of fatalities impact (e.g., Strickland et al. 2011; Rodrigues et al. 2015). Field work is typically conducted to count the number of carcasses observed over a certain period (e.g., a given season of the year) to allow the estimation of carcass persistence rates and searcher efficiency. These correction factors should be combined with the proportion of area effectively searched (including the number of wind turbines and the size of the area around each turbine searched, amongst other parameters) and should finally be used to adjust the number of carcasses found, in order to estimate the total number of animals killed.

The variety of existing estimators are distinguished by how the correction factors are determined, how they are incorporated into the estimators, and the technical and methodological assumptions. Over time, the number of available estimators has increased; thus, environmental specialists must select the method that best applies to their case in order to derive the associated correction factors and estimate fatality caused by the wind farm under study.

However, knowledge on the fatality caused by wind farms per se is not sufficient to determine the repercussion of the impact on the wildlife population's sustainability. The next step in the analysis is to estimate the population level impacts that such fatalities may have and by doing so, to determine the impact significance, as well as the adequate measures required to minimise it.

Previous publications have compared globally used fatality estimators regarding their mathematical and ecological assumptions (e.g., Bernardino et al. 2013; Huso et al. 2016). Our goal in this chapter is to provide an overview of the common fatality estimation practices used in monitoring programmes at wind farms in Portugal, as well as how population level impacts should be considered in impact evaluation.

7.1.1 Fatality Estimators: A Comparative Analysis

7.1.1.1 Commonly Used Estimators

We reviewed a total of 225 Portuguese fatality monitoring reports, published between 2005 and 2015 (Table 7.5, see Appendix), to determine: (i) if estimators have been used to adjust the number of carcasses observed, and estimate bat and bird fatality; (ii) which estimators are more commonly used; and (iii) if over the years companies have changed and/or updated the used estimators. The number of reports compiled for each year was, on average, 20 (min: 7; max: 31). In total, 31 reports had no records of bird or bat fatality, so they were excluded from the analysis. The final dataset included 194 reports from 14 different consultancy companies.

Most (91.8%) of the reports analysed used an estimator to adjust the observed fatality. The reports that did not (i.e. those that only reported the observed fatality)

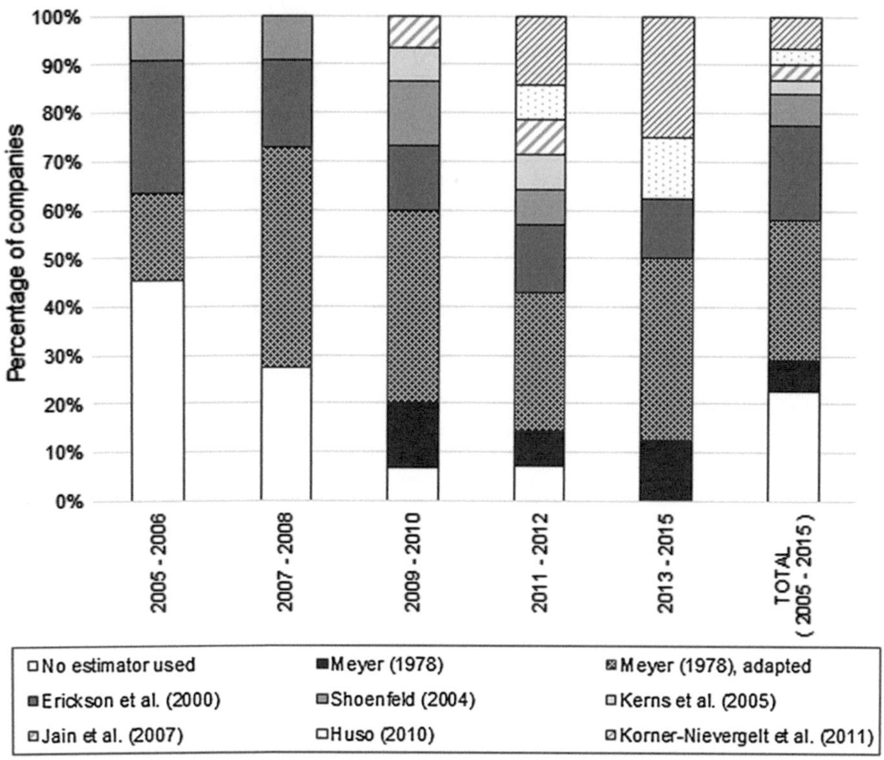

Fig. 7.1 Fatality estimators used in the Portuguese monitoring reports, between 2005 and 2015

were mainly published in the early years, meaning between 2005 and 2007 (n = 14; among the 16 produced in that period).

Between 2005 and 2015, eight different methods were used to obtain fatality estimates. Figure 7.1 shows which ones were most frequently adopted by the consultancy companies. An adapted version of the Meyer (1978) estimator was the most commonly used, although this was by companies responsible for a lower number of studies, and its formula can be simply defined as:

$$M = \frac{C}{a\,r\,d} \tag{7.1}$$

where M denotes the estimated fatality, C is the number of carcasses found during the carcass searches, a is the proportion of searched area, r is the carcass persistence empirical probability (i.e. the proportion of carcasses not being removed between two consecutive searches), and d corresponds to the searcher efficiency rate.

On fewer occasions, the original formula of Meyer (1978) was also used, which means that an extra adjustment factor is included in the formula to account for crippling (i.e. animals injured by the wind turbines that die outside of the search area and are undetected). However, the values used to adjust the crippling bias were based on literature from other types of infrastructures (e.g., power lines). Little is known about the mortality that is not detected by searchers due to crippling bias at wind turbines, since no research studies have been carried out on the topic. However, despite being a known factor able to increase the uncertainty in estimates (Smallwood et al. 2010), almost none of the estimators developed in the wind energy context include this adjustment factor.

Besides the crippling bias, there are other sources of uncertainty that are not accounted for by the simpler estimators but that the more complex estimators, developed over the last 15 years, have tried to accommodate. As an example, some of the now overcome limitations include: (i) the ability to estimate the search area considering that carcass density decreases with increasing distance from the wind turbine (Huso et al. 2015); and (ii) the consideration that carcasses may persist for several surveys and thus bleed-through is accommodated for—since searcher efficiency is almost never 100%, there is a possibility of a carcass not found during one search being found in a subsequent search (Bastos et al. 2013; Péron et al. 2013; Wolpert 2015). Three of the eight estimators used in the Portuguese reports already accommodate the assumption for bleed-through in their formulas, namely Shoenfeld (2004), Kerns et al. (2005) and Korner-Nievergelt et al. (2011). Although the first two are almost no longer used, Korner-Nievergelt et al. (2011) was among the estimators most used in the reviewed reports published between 2013 and 2015.

Bernardino et al. (2013), Korner-Nievergelt et al. (2015) and Huso et al. (2016) revised the differences between each estimator and their assumptions, both mathematically and ecologically, as well as some of their remaining limitations. Other authors, such as Huso (2011) and Korner-Nievergelt et al. (2011), conducted simulation studies to assess the performance of their newly proposed estimators, in comparison with other estimators previously used (e.g., Shoenfeld 2004). Although there is still no 'best' estimator that can be recommended regardless/despite of the circumstances of each wind farm, the simulation studies concluded that the latest estimators lead to better fatality estimates with lower associated uncertainty. However, not all consultancy companies have been able to follow up the research conducted on this topic and/or incorporate the use of new mathematical estimators in their reports.

In general, the estimators used by Portuguese specialists changed over the years. Several companies started using the simpler estimators such as the adapted version of Meyer (1978), Erickson et al. (2000) or Shoenfeld (2004) estimators in 2005–2006, and ended up using Huso (2011) and/or Korner-Nievergelt et al. (2011) estimators, in 2013–2015. This is to be expected, since with the release of new estimators dealing more effectively with known limitations, one would strive to produce more accurate fatality estimates. In some reports, one company used different estimators in order to have a better understanding of the ranges of possible fatality caused by the wind farms in study. Most Portuguese companies used at least three different estimators on the reports compiled over time, between 2005 and 2015. All these companies were responsible for a great number of reports, which may suggest they have more experience on the implementation of fatality

Table 7.1 Available platforms and software used to implement the fatality estimators most widely used by Portuguese consultancy companies

Fatality estimator (USGS software)[a]	Wildlife fatality estimator[b]	"Carcass" (R software package)[c]
Huso et al. (2011)	Jain et al. (2007), Huso et al. (2011), Korner-Nievergelt et al. (2011)	Shoelfeld (2004), Huso et al. (2011), Korner-Nievergelt et al. (2011)

[a]Fatality Estimator available at: https://pubs.er.usgs.gov/publication/ds729
[b]Wildlife Fatality Estimator available at: http://www.wildlifefatalityestimator.com/
[c]Package "carcass" for R Software available at: https://cran.r-project.org/web/packages/carcass/

monitoring studies and/or they were able to invest more time updating their methodological approaches. Nevertheless, there were also companies that despite conducting a high number of reports, used always the same estimator over the years (i.e. the Meyer (1978) estimator or its adapted version). None of the estimators published in the most recent years (e.g., Etterson 2013; Korner-Nievergelt et al. 2013; Huso et al. 2015) have been used by any of the consultancy companies.

Given the increasing complexity of the published estimators, software has been developed to help practitioners implementing these methods in their fatality monitoring studies. Examples include the web platform developed by Bio3 (www.wildlifefatiltyestimator.com), the USGS 'Fatality Estimator' software (Huso et al. 2012), and the R-package 'carcass' (R Core Team 2016; Korner-Nievergelt et al. 2015) (Table 7.1). However, currently there are no strict national or international guidelines on the 'best' estimator to use. Thus, practitioners can adopt the one they believe is best for their data or, in many cases, the one that is less demanding from a computational point of view or in terms of the statistical knowledge required for its application.

7.1.2 Recently Published Estimators

New approaches to fatality estimation have been developed, particularly from 2013 onwards. Their use is not yet current practice in Portugal, hence an integrated analysis of the requirements of such estimators and their possible utilisation by specialists conducting monitoring programmes is useful in order to improve the methods being implemented. We have analysed seven wind farm-caused fatality estimators published since 2013, as shown in Table 7.2. These estimators were evaluated for their mathematical and ecological assumptions, as well as advances in knowledge in relation to the commonly used estimators in Portugal.

Despite the method used, new estimators are similar in their approach to the fatality problematic: fatalities are modelled as a population with an occurrence within the wind farm, with its size being estimated depending on the correction factors and considered covariates (Table 7.2).

Table 7.2 Comparison of recent fatality estimation procedures published since 2013, with regards to their parameters and implementation methods

Variables considered	Bastos et al. (2013) (stochastic dynamic method)	Etterson (2013) (hidden Markov models)	Korner-Nievergelt et al. (2013) (Bayes theorem: adaptation of N-mixture model)		Péron et al. (2013) (superpopulation capture–recapture models)		Stevens and Dennis (2013) (state space model)	Huso et al. (2015) (Bayesian model-based approach)	Korner-Nievergelt et al. (2015) (adaptation of Korner-Nievergelt et al. (2011))	Wolpert (2015) (adaptation of four existing estimators)
			Observation sub-model	Collision sub-model	Multi-model approach (stochastic)	Regression approach (deterministic)				
Search interval	✓	✓	✓		✓	✓	✓	✓	✓	✓
Study length	✓	✓	✓		✓	✓	✓	✓		✓
Carcass persistence	✓	✓ (constant/variable with date/variable with age)	✓		✓ (variable with age)	✓	✓	✓	✓ (variable with date)	✓
Searcher efficiency	✓	✓ (constant/variable with date/variable with age)	✓		✓	✓	–	✓	✓	✓
Observed fatality	✓	✓	✓		(variable with age)	✓ (plus entry[a] probability)	✓	✓	✓	✓
Wind farm characteristics	✓ (number of wind turbines)	✓ (proportion of area searched)	✓ (proportion of area searched)	✓ (wind speed and rotor diameter)	✓ (proportion of area searched)	✓ (proportion of area searched)	–	✓ (proportion of area searched)	–	–
Ecological characteristics	–	–		✓ (Animal density)	–	–	–	–	–	–
Implementation software	Developed in Stella—model specific to	R package "carcass" (version without mark–recapture) and R	R package "carcass"		R package "carcassCMR"	R package "carcassCMR"	R (code made available	"Evidence of absence" software (Dalthorp et al. 2014)	R package "carcass"	R package "acmeR"

(continued)

Table 7.2 (continued)

Variables considered	Bastos et al. (2013) (stochastic dynamic method)	Etterson (2013) (hidden Markov models)	Korner-Nievergelt et al. (2013) (Bayes theorem: adaptation of N-mixture model)		Péron et al. (2013) (superpopulation capture–recapture models)		Stevens and Dennis (2013) (state space model)	Huso et al. (2015) (Bayesian model-based approach)	Korner-Nievergelt et al. (2015) (adaptation of Korner-Nievergelt et al. (2011))	Wolpert (2015) (adaptation of four existing estimators)
			Observation sub-model	Collision sub-model	Multi-model approach (stochastic)	Regression approach (deterministic)				
	the case study	and MATLAB code for all versions					by the authors)			

[a]Entry probability—the proportion of carcasses that enters the system (individuals die) between searches

Previous estimators were not necessarily flawed; however, they were simplistic representations of a complex network of interactions between wind farms operation and populations ecology. Some of the recently published estimators have such variables in consideration by allowing for differences in collision probability depending on the time of the year, since birds and bats vary behaviours and abundances seasonally. This is taken into consideration, for example, by Korner-Nievergelt et al. (2013), Stevens and Dennis (2013) and Péron et al. (2013), where it is considered that fatality rate is constant, though annual heterogeneity is allowed through the inclusion of covariates to introduce variations (e.g., activity peaks) through the study period. Indeed, Stevens and Dennis (2013) highlight the importance of modelling heterogeneity in carcass addition process (i.e. occurrence of collision mortality). Korner-Nievergelt et al. (2013) inclusively takes this thought process one step further by allowing the inclusion of data with regards to the characteristics of the community present and likely to suffer fatalities caused by wind turbines.

A novel addition of several new estimators is the ability to estimate mortality even in the absence of detected carcasses (i.e. Bastos et al. 2013; Huso et al. 2015; Wolpert 2015). These estimators, unlike other existing estimators allow the absence of observed mortality to be dealt with, producing estimations based on simulations or prior knowledge of the wind farm under study. Since this is the situation for many Portuguese wind farms (i.e. zero observed fatality annually), this prevents wrong interpretations of the meaning of the zero-estimated fatality by the decision-makers, by estimating fatalities that occurred but that were missed due to the bias factors.

This highlights the importance of the fatality estimation procedure in accommodating for different correction factors, such as carcass persistence probability, searcher efficiency and proportion of searched area. With regards to the recently published estimators, they have not dedicated as much focus on improving the estimation of correction factors. From the seven estimators analysed, there were only three implemented methods that differed from the traditional approach of estimating persistence probability and searcher efficiency. Bastos et al. (2013) used simulation techniques to reproduce the daily variability of carcass persistence and searcher efficiency. On the other hand, Péron et al. (2013) and Etterson (2013), in one of their estimator versions, resorted to mark–recapture models and respective transition matrices, which estimate the bias parameters from a unified analysis. Both methods proposed a new approach that could result in a lack of dedicated persistency and detectability trials requirement, given that specific field methods are implemented during the carcass search. To do so, carcass searches must be conducted by at least two observers, where each would record whether a given carcass (detected in previous searches) has been scavenged and searched for new carcasses. This approach provides simultaneous information on carcass persistency, searcher efficiency and observed fatalities. Similarly, the Wolpert (2015) estimator also provides a framework to estimate persistency and efficiency from the same data used to estimate fatality, by incorporating in the estimator itself the estimation of such bias factors from the same data. However, as Péron et al. (2013) and Etterson

(2013) point out, whether ancillary trial experiments need to be conducted depends of the specific characteristics of the study at stake and the availability of sufficient fatalities to obtain the necessary sample.

With regards to search area adjustments, most of the estimators analysed did not specifically consider any additional adjustments to the fatalities that may fall outside of the search area, besides the traditional approach as proposed by Meyer (1978). An exception was made by Etterson (2013), where a Horvitz and Thompson (1952) estimator was used, similarly to Huso (2011). This included areas not searched at all (turbines not monitored), and carcasses that fell outside of a searched area. As a result, Huso et al. (2015) considered that carcass arrival rate (i.e. the presence of carcasses within the search area) could be estimated using density-weighted proportions (Huso and Dalthorp 2014); and Korner-Nievergelt et al. (2015), included baseline data to estimate the parameter a (proportion of carcasses that fall outside the search area) to combine the spatial distribution of carcasses with the spatial distribution of the search area. A method to estimate a is not provided but instead a suggestion is made to incorporate these data into estimates.

One of the limitations identified for most of the common simple estimators used in Portugal, apart from Kerns et al. (2005) and Etterson (2013) estimators, was that carcasses not found in each search had zero probability of being detected in subsequent searches—the bleed-through effect. Ecologically, this may not be true, especially when the detection probability is much lower than one. In such cases, it should be considered that overlooked carcasses can be found in a subsequent search, if they remain at the site, without being removed (by scavengers or decomposition). This fact has been acknowledged by all estimators that have been evaluated, and is considered within their ecological assumptions. Similarly, requirements regarding search intervals have been relaxed for most estimators, which are now able to consider both regular and irregular search schemes. The only exceptions are those from Bastos et al. (2013) and Wolpert (2015), which require a regular interval scheme.

Considering the above, as well as the mathematical characteristics of the estimators, an analysis of the pros and cons of each is presented in Table 7.3. One feature of the analysis is that all estimators rely strongly on advanced statistical calculations, which reduce the scope of users who can implement these in regular monitoring programmes. An example of one of the most demanding estimators with regards to the calculations made is that from Bastos et al. (2013), who developed a model based on population dynamics of one species in a proprietary software offering. Nonetheless, many of them have developed and made available specific software or additional R-packages, which eases the comprehension and utilisation of these fatality estimators. Nonetheless, a basic understanding of the underlying mathematics and assumptions of each estimator is still essential to ensure the best results, as well as the use of R software© and the associated necessary packages. Thus, estimators considered to be of easier use are from Etterson (2013), Péron et al. (2013), Huso et al. (2015) and Wolpert (2015), which either have a dedicated software or an R-package with tutorials.

Table 7.3 Strong points, assumptions and potential limitations of using each analysed fatality estimator

Estimator		Limitations	Assumptions
Bastos et al. (2013)	• The dynamic nature of the estimator allows the actualization of important parameters, such as carcass persistence and searcher efficiency, throughout the simulation • The obtained algorithms are capable of estimating mortality even in the absence of detected carcasses	• The stochastic variables used in the simulation process use extreme information from European bat monitoring programmes and integrate them in two dynamic models, one explaining estimated and the other detected mortality	• Carcasses can be detected in subsequent searches (bleed-through). • It is assumed that no carcasses are present at the beginning of the study
Etterson (2013)	• Based on multi-event or hidden Markov models suitable under different experimental conditions. Scavenging and searcher efficiency can be estimated simultaneously from a unified analysis of a single trial under a variety of experimental conditions • Deals with irregular searches • The estimator has several versions that may be applied depending on the specific wind farm' characteristics. Some of them allow determination correction factors traditionally	• When observed mortality is 0, it yields 0 estimates • To adopt the mark-recapture alternative proposed, many animals killed are required to obtain enough carcasses detected by both observers to allow precise estimation of searcher efficiency and persistence without additional experimental trial data	• Fatalities are constant through the year

(continued)

Table 7.3 (continued)

Estimator		Limitations	Assumptions
Korner-Nievergelt et al. (2013)	• Model provides estimates per day unlike most models, which provide estimates for the whole period • Simulations show that this model could be used in very similar conditions to estimate fatality at wind turbines where no searches were conducted	• For the collision model to be accurate, a strong relationship must be found between animal density and the predictor environmental variable(s) • When observed mortality is 0, it yields 0 estimates	• To fit the model for prediction at new wind turbines the dataset needs to be large enough and the dataset/location would need to be the same
Péron et al. (2013)	• Super population models are modified to simultaneously analyse data from carcass surveys and from trial experiments • In theory carcass persistence probability and searcher efficiency can be estimated from the carcass search data alone using mark–recapture models • Estimates have less uncertainty due to the consideration of age and time variation in the parameters	• To adopt the mark–recapture alternative proposed, a large number of animals killed are required to obtain enough carcasses detected by both observers to allow precise estimation of searcher efficiency and persistence without additional experimental trial data • When observed mortality is 0, it yields 0 estimates	• Fatality rate is assumed to be constant (but option made available in the software for this rate to have a peak in the middle of the study period). • It is considered that carcasses may persist undetected for several intervals, and thus bleed-through is accommodated for
Stevens and Dennis (2013)	• This approach accommodates modelling heterogeneity (by allowing the inclusion of covariates) in the addition and removal processes using generalised regression, relaxing the assumption of mortality at a constant rate and allowing age-dependent persistence and searcher efficiency	• The fatality estimator does not account for searcher efficiency	• Carcass population change is modelled through a time-homogenous Markov process, which may be unrealistic for many applications (as suggested by Bispo et al. (2013)). One solution for relaxing this constraint is to use time-varying covariates in the regression functions

(continued)

Table 7.3 (continued)

Estimator		Limitations	Assumptions
Huso et al. (2015)	• Deal with cases where the observed mortality count is 0, and the probability of detection is lower than 1. Bayes' theorem is used to construct a posterior distribution for mortality, M (conditional on the observed mortality and estimated detectability). The distribution of M is then used to estimate two values that the authors consider important to conservation: the probability that M is below a predetermined limit and the upper bound (M*) of the 100(−α)% credible interval for M	• May produce results with a high confidence intervals in locations where no previous information exists, hence this may be of limited application is the first monitoring years	• The results may depend highly on the prior distribution chosen for mortality
Wolpert (2015)	• Introduces three new features to improve the reliability of mortality estimates: the diminishment of searcher efficiency as carcasses age; the reduced rate of scavenger removal as carcasses age • May deal with zero observations, by estimating one-sided confidence intervals for mean mortality similarly to Huso et al. (2015) • Includes interval (as well as point) mortality estimates	• No formal inclusion of a correction for the searchable area	• Some but not all carcasses present but undiscovered by the observer in one search may be discovered in a later search (flexible "bleed-through")

As previous studies have shown (e.g., Korner-Nievergelt et al. 2015; Huso et al. 2016), all recently published methods still have some limitations, while each possess some type of advance in knowledge with regards to overcoming a previously known flaw in fatality assessment. An estimator that applies to all possible situations does not exist, since factors influencing fatality monitoring design are far too many, as shown in Table 7.3. Therefore, it is advised that the assumptions and limitations of each estimator are taken into consideration upon use.

7.2 Population Level Impacts—Significance Estimation

Fatality estimators are used to evaluate the direct impacts caused by wind farms at the local level, i.e. caused by the development under assessment. Estimators developed so far are not focused on the impacts that wind farms can have on broader scale populations—aerial vertebrates specifically can have widespread and geographically disperse populations with large-scale movements between them. These populations can be affected by multiple-source fatalities (Erickson et al. 2014a), aggravated by the construction of wind farms in locations where the nucleus of these meta-populations exists, particularly in their 'source' locations. This question has been raised by many recent publications referring to impacts caused by wind energy on bird and bat populations (e.g., Bastos et al. 2015; Voigt et al. 2015; Zimmerling and Francis 2016).

Cumulative impacts
Cumulative impacts are "those that result from the successive, incremental, and/or combined effects of an action, project, or activity when added to other existing, planned, and/or reasonably anticipated future ones" (IFC 2013).

Such impacts and their consideration under the Environmental Impact Assessment (EIA) framework is recognised as the cumulative impact assessment. Though the specific evaluation of these impacts is regulated in the EIAs of several countries, it is only briefly referred to in the Portuguese legislative context (Decree-Law No. 131B/2013 of October 31st, amended by Decree-Law No. 179/2015 of August 27th) and the Guidelines for Impact Evaluation at Wind farms (APA 2010). These guidelines state that in the EIA report, other regional wind farm and power line monitoring reports should also be considered as well, in order to provide an analysis of the cumulative impacts based on the available knowledge. Though conducted with limitations, cumulative impact assessment in the Portuguese context has already led to the construction of wind farms being avoided due to the presence of other wind farms in the region (Amaral 2009).

A general methodology to assess cumulative impacts on birds has been proposed by Masden et al. (2010). This approach focuses on highlighting the actions, impacts and scales to be considered in this type of assessment, in order to decrease the uncertainty related to this evaluation type. As the authors point out, a general difficulty is expected in the data collection process since this type of evaluation pertains to impacts affecting the population scale, and therefore may include data gathering at broader (and perhaps the international) scale.

A more specific approach has been presented for the offshore context in relation to the seabirds' population level impacts (Poot et al. 2011). The methodology presented was focused on a species specific-level and included the construction of population models; scenario construction for wind farm development and calculation of models to determine increased mortality in sustainable levels. Finally, the authors compared the models obtained in order to provide a possible insight into the implementation of wind farms at the population level. However, even for this situation, the authors struggled to find data necessary to evaluate impacts at a broader scale, both on and offshore due to lack of information from nearby wind farms.

Onshore approaches conducted in the Mediterranean climate, including studies in Portugal, have tried to identify factors which constrain bird and bat fatalities at the landscape scale, hence considering factors that help to predict impacts at the population level (Table 7.4). This approach can be useful to determine impact probability at the regional scale before wind farm construction. In regard to this, tools have been published recently, including software to determine the population level impacts of wind farms in *Myotis* sp. bat populations (Erickson et al. 2014b). Similar to the study of Poot et al. (2011), this tool facilitate the evaluation methods using simulations of population-levels effects of disease and wind energy development. Santos et al. (2013) also used environmental features to explain and help predict locations with higher bat fatality probability in Portugal. Such features included forested areas and steep slopes, which support guidelines for wind farm placement best practices (Table 7.4). By predicting features with a higher probability of fatality occurrence, specialists may act on two fronts: on one hand wind turbine placement can be avoided; and on the other hand, they are able to predict that a higher impact on the population is expected from wind farms with turbines located within such risk features.

A different approach was implemented in Carrete et al. (2012) and Bastos et al. (2015) where individualised simulation models were used to estimate the evolution of the populations of target species (Egyptian Vulture *Neophron percnopterus* and Eurasian Skylark *Alauda arvensis*) considering the presence or foreseen expansion of wind farm installation in the species area in question, i.e. the Iberian Peninsula and Northern Portugal, respectively. Both studies demonstrated that populations of both species were expected to decrease due to additional fatality induced by the presence of wind energy facilities.

Table 7.4 Examples of studies conducted in Mediterranean climate that identify environmental factors explaining bird and bat presence at the regional and/or national scale

Reference	Geographic scope	Species/group	Environmental predictive impact variables
Bastos et al. (2015)	Northern Portugal	Skylark (*Alauda arvensis*)	Annual Precipitation % cover of urban areas % cover of agricultural areas % cover of annual crops % cover of broadleaf forests % cover of coniferous forests % cover of mixed forests % cover of pioneer mosaics Maximum percentage of burned area between 1990 and 2009 Density of local hydrographic network Altitude Local diversity of altitude types Number of patches present in each grid cell Mean gross primary productivity
Carrete et al. (2012)	Spain	Griffon Vulture (*Gyps fulvus*)	Bird aggregation in breeding season
Roscioni et al. (2013)	Central Italy	Lesser noctule (*Nyctalus leisleri*)	Widespread forests
Santos et al. (2013)	Portugal	Common pipistrelle (*Pipistrellus pipistrellus*) Lesser noctule (*Nyctalus leisleri*) Savi's pipistrelle (*Hypsugo savii*) Kuhl's pipistrelle (*Pipistrellus kuhlii*)	Forested areas (<5 km) Slopes steeper than 15° (<600 m) Precipitation ~600 mm Maximum temperatures ~18 °C Mean temperatures ~14 °C

An additional factor that is necessary to bear in mind while constructing models to understand a population sustainability level is that birds and bats are highly mobile and thus undertake migrations through the country or between countries. While such movements may be well studied for some species (e.g., Ramos Pereira et al. 2009; Rodrigues and Palmeirim 2008), other species' movements remain unknown, including not only the migration routes, but also the characteristics of the population that migrates. Such movements are a fundamental part of the population models, not only for wind farms located at the source and destination of these migrations, but also for all wind farms located in between the migration path, since during the migration period such wind farms will also contribute to an increased level of mortality in these populations.

The way forward in evaluating cumulative impacts caused by wind energy development is through utilisation of models to understand population dynamics and its interactions with wind farms. By creating a population dynamics model and adding fatalities caused by wind turbines as an additional source of mortality, environmental practitioners and decision makers could, in theory, understand the consequences of the construction and operation of multiple wind farms within the occupancy area of a given community or population. The objective of this assessment would be to understand the limits of additional mortality that a given population could sustain while still maintaining sustainable levels. This approach is particularly important for species of conservation concern, which already have depleted populations and any additional mortality caused by human-made infrastructures can lead to local and/or regional population extinctions. However, it is worth highlighting that wind farms are not the only source of bird and bat fatality, nor are they the most important for several species (Erickson et al. 2014a). Consequently, a unified database where information about fatalities found at any infrastructure could be a useful tool to determine with higher accuracy the state of the populations, as well as their current sustainability levels.

7.3 Discussion

Several Portuguese companies are still using obsolete estimators with numerous limitations, since they are easier to understand and/or to implement. Since estimators are constantly evolving and a point has not yet been reached where a global estimator has been found, it will be difficult for the competent national authorities to enforce the choice of an estimator. Therefore, the selection of the estimator is left to the consultant companies, and at most it is advised that multiple estimators are used to provide an insight into the possible range of estimated fatalities (Rodrigues et al. 2015). Several environmental practitioners, aiming to provide the best results throughout their monitoring programmes, have shown to be open to upgrading their practices over time by adopting new methods. They could make a leap forward and adjust their own methodologies in order to accommodate for model-based estimators. It is quite clear that most of the older and simplistic estimators have several ecological assumptions that are not realistic, which have been mostly overcome by several of the recently published model-based procedures (Bernardino et al. 2013; Huso et al. 2016).

The Korner-Nievergelt et al. (2013) fatality estimation procedure is somewhat more demanding in terms of the data required to determine cause of fatality, in comparison with other estimators, since it considers additional environmental and biological data to provide estimates with less uncertainty. Though this is line with the objectives of monitoring programmes, it can be quite costly due to the necessary equipment and the data analysis process, which are currently not part of the requirements of the best practice guidelines for wind farm assessment in Portugal. Thus, to our understanding, to implement any of the model-based methods, these would need to have a better cost-benefit relation than the currently used fatality estimation methods (e.g., by not requiring additional trials). Their utilisation would also be more widespread if a user-friendly software was available. Both Etterson (2013) and Péron et al. (2013) could be good candidates in these conditions' however, both Péron et al. (2013), as well as some of the estimator variations of Etterson (2013), use mark–recapture methods requiring that many animals are killed at a given wind farm in order to accurately estimate bias factors without additional experiments. This may be proven to be impractical at locations with low fatality numbers. Nonetheless, both are implemented through R software in already-built specific packages (Péron and Hines 2014; Korner-Nievergelt et al. 2015).

The lack of published best practice recommendations for fatality estimation implemented in Portugal is particularly important when specialists aim to understand and determine cumulative impact significance. Any study made to date to try to understand wind farm impacts at the national scale (see Marques et al., Chap. 1, Santos et al. 2013) could not use estimated fatality due to the heterogeneity of approaches used since 2005. This is not a problem restricted to the Portuguese reality but is universal to all countries. The development of a generalised estimator, able to standardise data analysis, reduce uncertainty and allow for comparable results across projects, both at the local and regional scale would be the next step towards evaluating cumulative mortality on wildlife populations. As previously mentioned, the first step towards the development of this generalised estimator could be a comparison study between the characteristics as well as to estimate accuracy (and associated uncertainty levels) of each existent estimator. This approach would allow for the identification of common flaws and to highlight the positive features, identifying the working areas to start from. Additionally, Huso et al. (2016) indicated that a specialised software should be developed to allow also non-statisticians to implement this generalised estimator.

Despite the lack of uniform fatality estimates, even if this data were made available and practitioners could estimate fatality caused by wind farms, a second barrier to determine cumulative impact significance would be the implementation of population level impact estimation. The next step in this research field could be by using population models to simulate impacts caused by multiple wind farms at the landscape level. However, several difficulties arise with this approach:

i. There are several species for which population dynamics are not well-known;
ii. If already known for a given species, population models are complex and difficult to implement without specific statistical knowledge;
iii. Simulation of impacts at the landscape level must be conducted per species or at least per groups of species with very similar life cycles and ecological requirement, i.e. to which the same model can be applied.

The mathematics behind modelling population dynamics is not simple, and besides BatTool, which already contains population models for *Myotis* sp., it is necessary to have knowledge on other software such as BUGS or R and population modelling techniques, to build the necessary models for other species, i.e. adapt the models to other species population dynamics, if known. Considering the difficulties that several environmental specialist experiences with the implementation of modelling techniques such as those used for fatality estimation, it is possible that the same situation may apply to the use of population dynamics models to estimate population level impacts. Thus, without a user-friendly software, containing information about the population parameters of species of interest, as well as the requirement from the decision makers to present this type of assessment, this field of study will remain lacking in Portuguese EIAs, with cumulative impact assessment being made qualitatively and superficially.

As a final remark, the next steps with regards to the assessment of impacts caused by wind farms is linked to the study of population dynamics. This is particularly important for species with a conservation status of concern, known to be killed at wind farms, since these are likely to be affected deeply at the population level, reaching risk levels for their sustainability. A second level of development with regards to this subject would be to set up a specific software or R-package to enable users with little knowledge on the mathematics behind it to estimate impact significance at the population level—such as the BatTool developed for *Myotis* sp. in the North American context (Erickson et al. 2014a). A desirable approach would be to develop this tool in such a way that it could be easily applied to several species or groups of species to optimise the resources for the consultancy companies conducting these studies. By enabling environmental practitioners with the tools to assess impact significance at the population level, using standardised methods (e.g., a single software for most species), a unified database could be generated at the national and/or international levels (Huso et al. 2016), easing exchange of information and allowing for proposed adequate mitigation measures, which is particularly important for migratory species.

Appendix

Table 7.5 List of consulted Portuguese monitoring reports, published between 2005 and 2015, respective wind farm location, faunal group assessed (birds and/or bats) and monitoring programme phase

Publication year	Authors	Title	Wind farm	Faunal group	Monitoring programme phase
2005	Plecotus (Silva B, Barreiro S, Cândido A, Alves P)	Parque Eólico de Chão Falcão. Monitorização de Quirópteros: Relatório 2—Ano 2005	Chão Falcão	Bats	2nd year (2005)
2005	Profico Ambiente	Monitorização ambiental da exploração da 2ª fase do Parque Eólico do Alvão—Relatório Final da Monitorização de Aves e Quirópteros para o Empreendimento Eólico de Alvadia	Alvão	Birds and bats	Final monitoring report
2005	Ecosfera	Projecto Eólico de Cinfães—Parques Eólicos de Alto do Talefe e Fonte da Quelha. Plano Geral de Monitorização—3º Relatório (2004 e primeiro trimestre de 2005) para a ENERNOVA	Alto do Talefe and Fonte da Quelha	Birds and bats	3rd year (2004/2005)
2005	Ecosfera	RECAPE da Ampliação do Parque Eólico de Pena Suar	Pena Suar	Birds and bats	Pre-construction
2006	Plecotus (Alves P, Silva B, Barreiro S)	Parques Eólicos na Serra dos Candeeiros. Monitorização de Quirópteros: Relatório 2 (Ano 2005)	Serra dos Candeeiros	Bats	2nd year (2005)
2006	ProSistemas	Plano de monitorização de fauna na área dos parques eólicos de Pinheiro e Cabril—2005 e Análise global dos resultados (2003–2005) para a Eólica da Cabreira	Pinheiro, Cabril and Cabreira	Birds and bats	Not defined (2005)
2006	Ecosfera	Projecto Eólico de Cinfães—Parques Eólicos de Alto do Talefe e Fonte da Quelha. Plano Geral de Monitorização	Alto do Talefe and Fonte da Quelha	Birds and bats	4th year (2005)

(continued)

Table 7.5 (continued)

Publication year	Authors	Title	Wind farm	Faunal group	Monitoring programme phase
		—4º Relatório (2005) para a ENERNOVA			
2006	Profico Ambiente	Parque Eólico do Outeiro—Programa de monitorização da mortalidade de aves e quirópteros—1º Relatório (Fevereiro–Março 2006)	Outeiro	Birds and bats	Not defined
2006	Profico Ambiente	Parque Eólico do Outeiro—Programa de monitorização da mortalidade de aves e quirópteros—2º Relatório (Março-Junho 2006)	Outeiro	Birds and bats	Not defined
2006	Profico Ambiente	Parque Eólico do Outeiro—Programa de monitorização da mortalidade de aves e quirópteros—3º Relatório (Junho, Julho e Setembro 2006)	Outeiro	Birds and bats	Not defined
2006	Strix	Plano de Monitorização: relatório de Verão—Parque Eólico do Portal da Freita (Serra do Marão), Ano 0—Trimestre 3—2006	Portal da Freita	Birds and bats	Pre-construction
2006	Strix	Monitorização das Comunidades de Aves e Quirópteros no Parque Eólico de Alagoa de Cima (Arcos de Valdevez), relatório de Fevereiro 2006	Alagoa de Cima	Birds and bats	Not defined
2006	Ecomind	Monitorização da mortalidade de aves e quirópteros no Parque Eólico da Lameira (Serra de Montemuro)—monitorização da mortalidade de aves e quirópteros (relatório de verão)	Lameira	Birds and bats	Not defined

(continued)

Table 7.5 (continued)

Publication year	Authors	Title	Wind farm	Faunal group	Monitoring programme phase
2006	Bio3	Monitorização da Comunidade de Aves do Parque Eólico de Chão-Falcão I. Relatório I (1º ano exploração)	Chão Falcão	Birds	1st year (2005)
2007	Plecotus (Hortêncio M, Barreiro S, Silva B, Alves P)	Parque Eólico do Caramulo: Monitorização de Quirópteros. Relatório 2—Ano 2006	Caramulo	Bats	2nd year (2006)
2007	Plecotus (Alves P, Geraldes P, Ferraz C, Silva B, Hortêncio M, Amorim F, Barreiro S)	Parque Eólico de Arada/Montemuro. Monitorização de Quirópteros: Relatório 2—Ano 2006	Arada/Montemuro	Bats	2nd year (2006)
2007	Plecotus (Barreiro S, Silva B, Alves P)	Parque Eólico da Serra dos Candeeiros (Candeeiros I e II): Monitorização de Quirópteros. Relatório 3—Ano 2006	Candeeiros	Bats	3rd year (2006)
2007	Plecotus (Silva B, Barreiro S, Alves P)	Parque Eólico de Chão Falcão. Monitorização de Quirópteros: Relatório 3—Ano 2006	Chão Falcão	Bats	3rd year (2006)
2007	Plecotus (Alves P, Geraldes P, Ferraz C, Hortêncio M, Silva B)	Parques Eólicos da Serra da Freita (Freita I e Freita II). Monitorização de Quirópteros: Relatório 2—Ano 2006	Serra da Freita	Bats	2nd year (2006)
2007	Plecotus (Lopes E, Alves P)	Sub-parques Eólicos de Mata Álvaro, Furnas e Seladolinho: Monitorização de Quirópteros. Relatório 2—Ano 2006	Mata Álvaro, Furnas e Seladolinho	Bats	2nd year (2006)
2007	Strix	Plano Especial de Monitorização de Quirópteros para a Serra do Marão—Parques eólicos de Penedo Ruivo, Seixinhos, Teixeiró e Mafómedes, Ano 1—2006	Marão	Bats	1st year (2006)

(continued)

Table 7.5 (continued)

Publication year	Authors	Title	Wind farm	Faunal group	Monitoring programme phase
2007	Profico Ambiente	Parque Eólico do Outeiro—Programa de monitorização da mortalidade de aves e quirópteros—4º Relatório (Setembro-Dezembro 2006)	Outeiro	Birds and bats	Not defined
2007	Ecosfera	Programa de Monitorização de Quirópteros do Projecto Eólico de Cinfães (Parques Eólicos de Fonte da Quelha e de Alto do Talefe)—Ano de 2006 para a ENERNOVA	Cinfães	Bats	Not defined
2007	Strix	Relatório Anual de Monitorização de Quirópteros do Parque Eólico da Videira, Ano de 2006	Videira	Bats	Not defined
2007	Ecosfera	Plano de Monitorização dos Quirópteros da Ampliação do Parque Eólico de Pena Suar—Fase de exploração—1º ano (2006) para a ENERNOVA	Pena Suar	Bats	1st year (2006)
2007	Strix	Monitorização das Comunidades de Aves e Quirópteros no Parque Eólico de Alagoa de Cima (Arcos de Valdevez), relatório da Primavera de 2006	Alagoa de Cima	Birds and bats	Not defined
2007	Strix	Monitorização das Comunidades de Aves e Quirópteros no Parque Eólico de Alagoa de Cima (Arcos de Valdevez), relatório do Verão de 2006	Alagoa de Cima	Birds and bats	Not defined
2007	Strix	Monitorização das Comunidades de Aves e Quirópteros no Parque Eólico de	Alagoa de Cima	Birds and bats	Not defined

(continued)

Table 7.5 (continued)

Publication year	Authors	Title	Wind farm	Faunal group	Monitoring programme phase
		Alagoa de Cima (Arcos de Valdevez), relatório do Outono de 2006			
2007	Strix	Plano Especial de Monitorização de Quirópteros para a Serra do Marão—Parques eólicos de Penedo Ruivo, Seixinhos, Teixeiró e Mafómedes, Ano 2–2007	Marão	Bats	2nd year (2007)
2007	ProSistemas	Plano de monitorização da fauna. Parque Eólico da Arga—2006/2007 para os Empreendimentos Eólicos da Espiga	Arga	Birds and bats	Not defined
2007	Colmus	Relatório de monitorização da fauna e paisagem do Parque Eólico das Terras Altas de Fafe (Setembro 2005—Dezembro 2006)	Terras Altas de Fafe	Birds and bats	Not defined
2007	Strix	Monitorização das Comunidades de Aves e Quirópteros no Parque Eólico de Alagoa de Cima (Arcos de Valdevez), relatório do Inverno 2007	Alagoa de Cima	Birds and bats	Not defined
2007	LEA	Monitorização da mortalidade de aves e quirópteros no Parque Eólico da Lameira. Relatório Final 2006/2007	Lameira	Birds and bats	Final monitoring report
2007	Bio3	Monitorização da Comunidade de Aves do Parque Eólico da Serra de Candeeiros (Candeeiros I e II). Relatório 2 (Ano 2006)	Candeeiros	Birds	2nd year (2006)

(continued)

Table 7.5 (continued)

Publication year	Authors	Title	Wind farm	Faunal group	Monitoring programme phase
2007	Bio3	Monitorização da Comunidade de Aves do Parque Eólico do Caramulo. Relatório II (Ano 2006)	Caramulo	Birds	2nd year (2006)
2007	Bio3	Monitorização da Comunidade de Aves do Parque Eólico de Chão-Falcão I. Relatório 2 (Fase de exploração - Ano 2006)	Chão Falcão	Birds	2nd year (2006)
2007	Bio3	Monitorização da Comunidade de Aves do Parque Eólico de Pinhal Interior. Relatório II (Ano 2006)	Pinhal Interior	Birds	2nd year (2006)
2008	Plecotus (Lopes S, Silva B, Alves P)	Sub-Parques Eólicos de Proença I e II. Monitorização de Quirópteros: Relatório 1—Ano 2006	Proença I and II	Bats	1st year (2006)
2008	Strix	Monitorização das Comunidades de Aves e Quirópteros no Parque Eólico de Caravelas (Vila Real)—Inverno de 2006	Caravelas	Birds and bats	Not defined
2008	Strix	Monitorização das Comunidades de Aves e Quirópteros no Parque Eólico de Caravelas (Vila Real)—Primavera de 2006	Caravelas	Birds and bats	Not defined
2008	Strix	Monitorização das Comunidades de Aves e Quirópteros no Parque Eólico de Caravelas (Vila Real)—Verão de 2006	Caravelas	Birds and bats	Not defined
2008	Strix	Relatório Anual de Monitorização de Quirópteros do Parque Eólico da Videira, Ano de 2007	Videira	Bats	Not defined

(continued)

Table 7.5 (continued)

Publication year	Authors	Title	Wind farm	Faunal group	Monitoring programme phase
2008	Plecotus (Silva B, Barreiro S, Hortêncio M, Alves P)	Parque Eólico do Caramulo: Monitorização de Quirópteros. Relatório 3—Ano 2007	Caramulo	Bats	3rd year (2007)
2008	Ecosfera	Plano de Monitorização dos Quirópteros da Ampliação do Parque Eólico de Pena Suar—Relatório de conclusão do plano, de 2004 a 2007 para a ENERNOVA	Pena Suar	Bats	Final monitoring report (2004–2007)
2008	Strix	Monitorização das Comunidades de Aves e Quirópteros no Parque Eólico de Alagoa de Cima (Arcos de Valdevez)—Primavera de 2007	Alagoa de Cima	Birds and bats	Not defined
2008	Strix	Monitorização das Comunidades de Aves e Quirópteros no Parque Eólico de Alagoa de Cima (Arcos de Valdevez)—Verão de 2007	Alagoa de Cima	Birds and bats	Not defined
2008	Strix	Monitorização das Comunidades de Aves e Quirópteros no Parque Eólico de Alagoa de Cima (Arcos de Valdevez)—Outono de 2007	Alagoa de Cima	Birds and bats	Not defined
2008	Strix	Relatório Anual de Monitorização de Quirópteros do Parque Eólico da Videira, Ano de 2008	Videira	Bats	Not defined
2008	LEA (Cabral JA, Barros P, Silva-Santos P, Silva C, Correia J, Gonçalves R, Braz L)	Programa de Monitorização da Actividade e Mortalidade de Quirópteros—Parque Eólico do Outeiro (Relatório Final—Fase de exploração)	Outeiro	Bats	Final monitoring report

(continued)

Table 7.5 (continued)

Publication year	Authors	Title	Wind farm	Faunal group	Monitoring programme phase
2008	Strix	Monitorização das Comunidades de Aves e Quirópteros no Parque Eólico de Alagoa de Cima (Arcos de Valdevez)—Inverno de 2008	Alagoa de Cima	Birds and bats	Not defined
2008	Colmus	Monitorização da avifauna e dos quirópteros—Fevereiro a Outubro 2008. Parque Eólico do Alto do Monção para a Eonergi	Alto do Monção	Birds and bats	Not defined
2008	Colmus (Duarte J, Moreira JP, Sousa V, Pereira F, Rodrigues P)	Monitorização da Avifauna, dos Quirópteros, e do Ruído Ambiente no Parque Eólico da Serra do Ralo—Relatório Final do Primeiro Ano de Exploração	Serra do Ralo	Birds and bats	1st year (2007)
2008	PROCESL	Parque Eólico de Pracana (Mação). Monitorização avifauna e quirópteros. Relatório final (2006 e 2007)	Pracana	Birds and bats	Final monitoring report
2008	Bio3	Monitorização da Comunidade de Aves do Parque Eólico do Caramulo. Relatório III (Fase II - 2º ano de exploração)	Caramulo	Birds	2nd year (2007)
2008	Bio3	Monitorização da Comunidade de Aves do Parque Eólico da Gardunha. Relatório II (Ano 2007—Fase de construção/início da Exploração)	Gardunha	Birds	Construction/1st year (2007)
2008	Bio3	Monitorização da Comunidade de Aves do Parque Eólico de Pinhal Interior. Relatório III (Ano 2007)	Pinhal Interior	Birds	3rd year (2007)

(continued)

Table 7.5 (continued)

Publication year	Authors	Title	Wind farm	Faunal group	Monitoring programme phase
2009	Plecotus (Hortêncio M, Barreiro S, Silva B, Alves P)	Monitorização de Morcegos no Parque Eólico de Chão Falcão. Relatório 4—Ano 2007	Chão Falcão	Bats	4th year (2007)
2009	Plecotus (Alves P, Barreiro S, Silva B)	Parque Eólico da Gardunha: Monitorização de Quirópteros. Relatório 2—Ano 2007	Gardunha	Bats	2nd year (2007)
2009	Plecotus (Alves P, Barreiro S, Silva B)	Parque Eólico da Lousã I: Monitorização de Quirópteros. Relatório 2—Ano 2007	Lousã I	Bats	2nd year (2007)
2009	Plecotus (Lopes S, Silva B, Alves P)	Sub-Parque Eólico do Moradal. Monitorização de Quirópteros: Relatório 2—Ano 2007	Moradal	Bats	2nd year (2007)
2009	Plecotus (Alves P, Lopes E, Barreiro B, Silva B)	Sub-parques eólicos de Mata-Álvaro, Furnas e Seladolinho. Monitorização de quirópteros. Relatório 3—Ano 2007 (relatório final)	Mata Álvaro, Furnas e Seladolinho	Bats	3rd year (2007)
2009	ProSistemas	Monitorização de quirópteros do Parque Eólico da Serra dos Candeeiros (Candeeiros I e II) para a Companhia de Energias Renováveis da Serra dos Candeeiros lda	Candeeiros	Bats	Not defined
2009	Strix	Plano Especial de Monitorização de Quirópteros para a Serra do Marão—Parques eólicos de Penedo Ruivo, Seixinhos, Teixeiró e Mafómedes, Ano 3—2008	Marão	Bats	3rd year (2008)

(continued)

Table 7.5 (continued)

Publication year	Authors	Title	Wind farm	Faunal group	Monitoring programme phase
2009	Profico Ambiente/Bio3	Monitorização da comunidade de aves e quirópteros no Parque Eólico da Guarda. Relatório II (Ano 2008)	Guarda	Birds and bats	2nd year (2008)
2009	Plecotus (Barreiro S, Silva B, Alves P)	Parque Eólico de Mosqueiros I: Monitorização de Quirópteros. Relatório 2—Ano 2008	Mosqueiros I	Bats	2nd year (2008)
2009	PROCESL	Monitorização de quirópteros 2008—Parque Eólico do Alto Minho I (Sub-parques Picos, Alto do Corisco e Santo António), para a VentoMinho	Alto Minho I	Bats	Not defined
2009	Colmus (Duarte J, Moreira JP, Cabral JA, Barros P, Silva C, Gonçalves H)	Relatório anual de monitorização da avifauna e dos quirópteros do Parque Eólico do Alto do Monção (Fevereiro de 2008 a Outubro de 2009)	Monção	Birds and bats	Not defined
2009	ProSistemas	Programa de monitorização ambiental do parque eólico de Arga (2007–2008) para os Empreendimentos Eólicos da Espiga	Arga	Birds and bats	Not defined
2009	AgriPro Ambiente	Avaliação do impacte nos quirópteros do projecto de execução do parque eólico da Serra da Alvoaça (Nov 2007–2008) para a EDP	Serra da Alvoaça	Bats	Not defined
2009	LEA (Duarte J, Moreira P, Sousa V, Cabral JA, Barros P, Silva-Santos P, Silva C, Correia J, Gonçalves R, Braz L)	Relatório de Monitorização da Avifauna, Quirópteros e Factores Socioeconómicos do Parque Eólico da Serra do Ralo (PESR)—Março de 2007 a Fevereiro de 2009, Relatório Final	Serra do Ralo	Birds and bats	Not defined

(continued)

Table 7.5 (continued)

Publication year	Authors	Title	Wind farm	Faunal group	Monitoring programme phase
2009	ProSistemas	Eólica da Arada, S.A. Monitorização de Quirópteros no Parque Eólico de Arada-Montemuro, Relatório 2	Arada/Montemuro	Bats	2nd year (2008)
2009	Ecosfera	Programa de estudos e monitorização da conservação da natureza do parque eólico de Negrelo e Guilhado. Monitorização de Quirópteros—Actividade e Mortalidade na Área do Parque Eólico de Negrelo e Guilhado. Fase de exploração—Ano de 2009	Negrelo e Guilhado	Bats	Not defined
2009	Plecotus	Parque Eólico de Mosqueiros I. Monitorização de quirópteros. Relatório 2—Ano 2009	Mosqueiros I	Bats	2nd year (2009)
2009	ProSistemas	Parque Eólico de Arga. Programa de monitorização ambiental - Relatório final da fase II—Ano 4 (2009)	Arga	Birds and bats	4th year (2009)
2009	Naturiberica	Monitorização do Parque Eólico de Alto da Folgorosa. Avifauna e Quirópteros. Relatório Anual	Alto da Folgorosa	Birds and bats	Not defined
2009	Bio3	Monitorização da Comunidade de Aves do Parque Eólico da Serra de Candeeiros (Candeeiros I e II). Relatório 3 (fase de exploração—Ano 2007)	Candeeiros	Birds	3rd year (2007)
2009	Bio3	Monitorização da Comunidade de Aves do Parque Eólico de Chão Falcão I. Relatório 3 (fase de exploração—Ano 2007)	Chão Falcão	Birds	3rd year (2007)

(continued)

Table 7.5 (continued)

Publication year	Authors	Title	Wind farm	Faunal group	Monitoring programme phase
2009	Bio3	Monitorização da Comunidade de Aves do Parque Eólico da Lousã I. Relatório III (Ano 2008—fase de exploração)	Lousã I	Birds	3rd year (2008)
2010	Plecotus (Alves P, Lopes E, Barreiro B, Silva B)	Sub-parques Eólicos de Proença I e II. Monitorização de Quirópteros. Relatório 2—Ano 2007	Proença I and II	Bats	2nd year (2007)
2010	Plecotus (Alves P, Geraldes P, Silva B, Amorim F, Ferraz C, Hortênsio M, Lopes E)	Parques Eólicos da Serra da Freita (Freita I e Freita II). Monitorização de quirópteros, relatório 3—2007	Serra da Freita	Bats	3rd year (2007)
2010	Plecotus (Barreiro S, Silva B, Alves P)	Parque Eólico do Caramulo—Monitorização de quirópteros. Relatório 4—2008 (Relatório final)	Caramulo	Bats	Final monitoring report
2010	Plecotus (Alves P, Lopes S, Silva B, Gonçalves R)	Sub-parque eólico do Moradal. Monitorização de quirópteros. Relatório 3—2008 (Relatório final)	Moradal	Bats	Final monitoring report
2010	Plecotus (Silva B, Barreiro S, Lopes E, Alves P)	Parque Eólico da Gardunha - Monitorização de Quirópteros. Relatório 3—Ano 2008	Gardunha	Bats	3rd year (2008)
2010	ProSistemas	Monitorização de Quirópteros no Parque Eólico de Serra de Sicó. Fase de Exploração—Relatório 1	Serra de Sicó	Bats	1st year (2009)
2010	Profico Ambiente/Plecotus	Monitorização dos quirópteros do Parque Eólico de Ribabelide—Relatório Final de Monitorização 2009	Ribabelide	Bats	Final monitoring report
2010	Profico Ambiente/Plecotus	Monitorização dos quirópteros do Parque Eólico da Lagoa de D. João e Feirão—Relatório Final de Monitorização 2009	Lagoa de D. João e Feirão	Bats	Not defined

(continued)

Table 7.5 (continued)

Publication year	Authors	Title	Wind farm	Faunal group	Monitoring programme phase
2010	Procesl	Monitorização da Comunidade de Quirópteros do Parque Eólico do Alto Minho I: Sub-Parques Eólicos de Picos, Alto Corisco e Santo António. Relatório III (Fase de Exploração—Ano 2009)	Alto Minho I	Bats	3rd year (2009)
2010	Ecosativa	Monitorização de Avifauna e Quirópteros do Parque Eólico de Madrinha. 7º Relatório de Progresso para a empresa ENERNOVA—Novas Energias, S.A.	Madrinha	Birds and bats	7th year (2009)
2010	LEA	Relatório de Monitorização da Actividade e Mortalidade de Quirópteros no Parque Eólico de Mafômedes. Relatório Final (Fase de Exploração) elaborado para a EnergieKontor	Mafômedes	Bats	Final monitoring report
2010	LEA	Relatório de Monitorização da Actividade e Mortalidade de Quirópteros nos Parques Eólicos de Penedo Ruivo e Seixinhos. Relatório Final (Fase de Exploração—Ano IV) elaborado para a EnergieKontor	Penedo Ruivo e Seixinhos	Bats	Final monitoring report
2010	Mae d´Agua	Parque Eólico da Videira—monitorização dos vertebrados voadores; efeito de exclusão/mortalidade 2010	Videira	Birds and bats	Not defined

(continued)

Table 7.5 (continued)

Publication year	Authors	Title	Wind farm	Faunal group	Monitoring programme phase
2010	ProSistemas	Parque Eólico de Arada-ontemuro: monitorização de quirópteros, fase de exploração 2º ano, relatório 3	Arada/Montemuro	Bats	2nd year (2009)
2010	Naturiberica	Monitorização do Parque Eólico de Alto da Folgorosa, relatório final 2010	Alto da Folgorosa	Birds and bats	Not defined
2010	ProSistemas	Parques Eólicos de Bustelo e Cinfães: Monitorização de Quirópteros, Fase de Exploração – 1º Ano	Bustelo and Cinfães	Bats	1st year (2009)
2010	Ecosativa	Monitorização do Parque Eólico da Serra do Alvão e Linha de Transporte de Energia. 1º Relatório Anual para a sociedade Parque Eólico da Serra do Alvão, S.A.	Serra do Alvão	Birds and bats	1st year (2009)
2010	Ecosativa	Monitorização de avifauna e quirópteros do Parque Eólico de Cadafaz II. 4º Relatório de Progresso para a empresa ENEOP2	Cadafaz II	Birds and bats	4th year (2009)
2010	PROCESL	Parque Eólico da Serra Alta - Monitorização da Avifauna. Fase de Exploração (Ano I - 2009/2010)	Serra Alta	Birds	1st year (2009/2010)
2010	Bio3	Monitorização da comunidade de aves no Sub-Parque Eólico do Mastro. Relatório Final	Mastro	Birds	Final monitoring Report
2010	Bio3	Monitorização da comunidade de aves no Parque Eólico da Gardunha. Relatório final (2006 a 2009)	Gardunha	Birds	Final monitoring report

(continued)

Table 7.5 (continued)

Publication year	Authors	Title	Wind farm	Faunal group	Monitoring programme phase
2010	Bio3	Monitorização da comunidade de aves e quirópteros no Parque Eólico da Guarda. Relatório III—Ano 2009 (Final)	Guarda	Birds and bats	Final monitoring report
2011	Prosistemas/Plecotus	Parque Eólico do Alto Minho I—subparque de Bustavade-Mendoiro, monitorização de quirópteros, relatório II, fase de exploração ano 1	Alto Minho I	Bats	1st year (2010)
2011	Plecotus (Alves P, Lopes S, Silva B)	Sub-parques de Proença I e II. Monitorização de quirópteros, relatório 3—2008	Proença I and II	Bats	3rd year (2008)
2011	Plecotus (Alves P, Silva B, Barreiro S)	Parque Eólico da Lousã I. Monitorização de quirópteros. Relatório 3—Ano 2008	Lousã I	Bats	3rd year (2008)
2011	Plecotus	Monitorização dos quirópteros do Parque Eólico de Ribabelide—Relatório Final de Monitorização 09/10	Ribabelide	Bats	Final monitoring Report
2011	ProSistemas	Parque Eólico do Alto Minho I - Subparque eólico de Bustavade-Mendoiro. Monitorização de quirópteros - relatório III, fase de exploração—Ano 2	Bustavade-Mendoiro	Bats	2nd year (2010)
2011	Plecotus (Alves P, Silva B, Barreiro S)	Parque eólico de Mosqueiros II. Monitorização de quirópteros. Relatório 2—Ano 2009	Mosqueiros II	Bats	2nd year (2009)
2011	Biota	Estudo de monitorização das comunidades de aves, lobo, morcegos, flora, vegetação e habitats do Parque Eólico e Linha Eléctrica do Sabugal.	Sabugal	Birds and bats	Not defined

(continued)

Table 7.5 (continued)

Publication year	Authors	Title	Wind farm	Faunal group	Monitoring programme phase
2011	Profico Ambiente/Plecotus	Monitorização dos quirópteros do Parque Eólico da Lagoa de D. João e Feirão—Relatório Final de Monitorização 2010	Lagoa de D. João e Feirão	Bats	Final monitoring report
2011	ProSistemas	Parque Eólico de Serra do Barroso III: Monitorização de Quirópteros, Fase de Exploração – 1º Ano	Barroso III	Bats	1st year (2010)
2011	ProSistemas	Parque Eólico de Vila Nova II: Monitorização de Quirópteros, Fase de Exploração – 1º Ano	Vila Nova II	Bats	1st year (2010)
2011	LEA	Relatório de Monitorização da Actividade e Mortalidade de Quirópteros no ParqueEólico de Mafômedes. Relatório Anual (Ano 3—fase de exploração) elaborado para a EnergieKontor	Mafômedes	Bats	3rd year (2010)
2011	LEA	Relatório de Monitorização da Actividade e Mortalidade de Quirópteros nos Parques Eólicos de Penedo Ruivo e Seixinhos. Relatório Final (Ano 5. Fase de exploração) elaborado para a EnergieKontor	Penedo Ruivo e Seixinhos	Bats	5th year (2010)
2011	ProSistemas	Parque Eólico de Serra do Sicó. Monitorização de quirópteros—relatório 2, fase de exploração	Serra de Sicó	Bats	2nd year (2010)
2011	LEA	Programa de estudos e monitorização da conservação da natureza do parque	Negrelo e Guilhado	Bats	Not defined

(continued)

Table 7.5 (continued)

Publication year	Authors	Title	Wind farm	Faunal group	Monitoring programme phase
		eólico de Negrelo e Guilhado. Monitorização de Quirópteros—Actividade e Mortalidade na Área do Parque Eólico de Negrelo e Guilhado. Fase de exploração—Ano de 2010			
2011	Strix	Relatório Anual de Monitorização de Quirópteros do Parque Eólico do Barão de São João, Ano de 2010	Barão de São João	Bats	Not defined
2011	Procesl	Parque Eólico do Bairro (Pias Longas)—monitorização de quirópteros. Relatório 2 (1° ano da fase de exploração—2009/2010)	Bairro	Bats	1st year (2009/2010)
2011	Bio3	Parque Eólico de Chão Falcão III—Monitorização da comunidade de quirópteros. Relatório 1 (Fase de exploração)	Chão Falcão III	Bats	1st year (2010)
2011	ProSistemas	Parque Eólico de S. João I. Monitorização de quirópteros—relatório 1, fase de exploração	S. João I	Bats	1st year (2010)
2011	ProSistemas	Parque Eólico de S. João I. Monitorização de quirópteros—relatório 2, fase de exploração	S. João I	Bats	2nd year (2010)
2011	Bio3	Monitorização das comunidades de aves e quirópteros. Relatório final (2007 a 2010). Parque Eólico da Serra do Mú	Serra do Mú	Birds and bats	Final monitoring report (2007/2010)

(continued)

Table 7.5 (continued)

Publication year	Authors	Title	Wind farm	Faunal group	Monitoring programme phase
2011	Ecosativa	Monitorização de avifauna e quirópteros do Parque Eólico de Cadafaz II. 5º Relatório de Progresso para a empresa ENEOP2 (Outubro 2010—Junho 2011)	Cadafaz II	Birds and bats	Not defined
2011	Ecosativa	Monitorização de avifauna e quirópteros do Parque Eólico de Cadafaz II. 6º Relatório de Progresso para a empresa ENEOP2 (Julho—Setembro 2011)	Cadafaz II	Birds and bats	Not defined
2011	Ecosativa	Plano de Monitorização de avifauna e quirópteros do Parque Eólico de Fonte da Mesa II. Fase de exploração. 1º Relatório Anual para a empresa Eólica do Alto Douro	Fonte da Mesa II	Birds and bats	1st year (2010)
2011	Ecosativa	Plano de Monitorização de avifauna e quirópteros do Parque Eólico de Balocas e Linha de Transporte de Energia. Fase de exploração. 1º Relatório Anual para a empresa Eólica das Serras das Beiras, S. A.	Balocas	Birds and bats	1st year (2010)
2011	Ecosativa	Plano de Monitorização de avifauna, quirópteros, flora e vegetação do Parque Eólico de Açor II. Fase de exploração. 1º Relatório Semestral para a empresa Eólica das Serras das Beiras, S.A.	Açor II	Birds and bats	Not defined
2011	ProSistemas	Monitorização de quirópteros—Parques Eólicos de Bustelo e Cinfães. Fase de exploração—2º ano	Bustelo and Cinfães	Bats	2nd year (2010)

(continued)

Table 7.5 (continued)

Publication year	Authors	Title	Wind farm	Faunal group	Monitoring programme phase
2011	ProSistemas	Monitorização de quirópteros—Parque Eólico de Arada-Montemuro. Fase de exploração—3º ano	Arada/Montemuro	Bats	3rd year (2010)
2011	Ecosativa	Monitorização de Avifauna e Quirópteros do Parque Eólico de Madrinha. Relatório final para a empresa ENERNOVA –Novas Energias, S.A.	Madrinha	Birds and bats	Final monitoring report
2011	Ecosativa	Plano de Monitorização de avifauna e quirópteros do Parque Eólico de Alto Arganil. Fase de Exploração. 1º Relatório anual para a empresa EDP Renováveis Portugal S.A.	Alto Arganil	Birds and bats	1st year (2010)
2011	Ecosativa	Monitorização do Parque Eólico da Serra do Alvão e linha de transporte de energia. Fase de exploração. 2º Relatório anual para a empresa Sociedade Parque Eólico da Serra do Alvão, S.A.	Serra do Alvão	Birds and bats	Not defined
2011	LEA	Relatório de Monitorização da Actividade e Mortalidade de Quirópteros no Parque Eólico do Sobrado. Relatório Final (Ano 2—Fase de Exploração) elaborado para a EnergieKontor	Sobrado	Bats	2nd year (2010)
2011	LEA	Relatório de Monitorização da Atividade e Mortalidade de Quirópteros no Parque Eólico do Sobrado. Relatório Final (Ano 3. Fase de exploração) elaborado para a EnergieKontor	Sobrado	Bats	3rd year (2010)

(continued)

Table 7.5 (continued)

Publication year	Authors	Title	Wind farm	Faunal group	Monitoring programme phase
2011	PROCESL	Parque Eólico da Serra Alta—Monitorização da Avifauna. Fase de Exploração (Ano II - 2011)	Serra Alta	Birds	2nd year (2011)
2011	Ecosativa	Monitorização de avifauna, quirópteros, flora e vegetação do Parque Eólico do Toutiço e respectivas linhas eléctricas. Fase de exploração. 1º Relatório Anual	Toutiço	Birds and bats	1st year (2010)
2011	Naturiberica	Monitorização Ambiental do Parque Eólico de Alto da Folgorosa—Fase de exploração. Avifauna, Morcegos, Flora e Vegetação	Alto da Folgorosa	Birds and bats	Not defined
2011	Bio3	Monitorização da comunidade de aves no Parque Eólico de S. Macário I. Relatório 5 (Fase de exploração—3º Ano). Relatório Final	S. Macário I	Birds	3rd year (2010)
2011	Bio3	Monitorização da Comunidade de Aves do Parque Eólico de Chão-Falcão III. Relatório III (Fase de exploração)	Chão Falcão III	Birds	3rd year (2010)
2012	Plecotus	Parque Eólico da Serra dos Candeeiros (Candeeiros I e II). Monitorização de Quirópteros. Relatório 05	Candeeiros	Bats	5th year (2008)
2012	Plecotus (Alves P, Silva B, Barreiro S)	Parque Eólico de Candeeiros. Monitorização de Quirópteros. Relatório 6—Ano 2009	Candeeiros	Bats	6th year (2009)
2012	ProSistemas	Parque Eólico de Vila Nova II: Monitorização de Quirópteros, Fase de Exploração—2º Ano	Vila Nova II	Bats	2nd year (2011)

(continued)

Table 7.5 (continued)

Publication year	Authors	Title	Wind farm	Faunal group	Monitoring programme phase
2012	BIOTA	Parque Eólico do Sabugal—Estudo de monitorização das comunidades de aves, lobo, morcegos, flora e vegetação. Relatório Final. Fase de exploração—3º Ano	Sabugal	Birds and bats	3rd year (2011)
2012	PROCESL	Monitorização de quirópteros do Parque Eólico de Alvaiázere. Relatório anual (ano 1 da fase de exploração—2011)	Alvaiázere	Bats	1st year (2011)
2012	Strix	Relatório Anual de Monitorização de Quirópteros do Parque Eólico do Barão de São João, Ano de 2011	Barão de São João	Bats	Not defined
2012	ProSistemas	Parque Eólico de Serra de Sicó. Monitorização de quirópteros. Fase de exploração—Relatório 3	Serra de Sicó	Bats	3rd year (2011)
2012	PROCESL	Parque Eólico da Raia (Sub-Parques de Benespera, Pousafoles e S. Cornélio)—monitorização de quirópteros. Relatório 3 (1º ano da fase de exploração—2011)	Raia	Bats	1st year (2011)
2012	LEA	Relatório de Monitorização da Atividade e Mortalidade de Quirópteros nos Parques Eólicos de Penedo Ruivo e Seixinhos. Relatório Final (Ano 6. Fase de exploração—Ano 2011) elaborado para a EnergieKontor.	Penedo Ruivo e Seixinhos	Bats	6th year (2011)
2012	LEA	Relatório de Monitorização da Atividade e Mortalidade de Quirópteros no Parque Eólico de Negrelo e Guilhado. Relatório Anual (Ano 3. Fase de exploração—	Negrelo e Guilhado	Bats	3rd year (2011)

(continued)

Table 7.5 (continued)

Publication year	Authors	Title	Wind farm	Faunal group	Monitoring programme phase
		2011) elaborado para a Ecosfera—Consultoria Ambiental, Lda			
2012	Noctula	Monitorização dos sistemas ecológicos na área do Parque Eólico de Coentral-Safra. Fase de exploração.	Coentral-Safra	Birds and bats	Not defined
2012	ProSistemas/Plecotus	Monitorização de quirópteros—Parques Eólicos de S. Pedro, de Casais e de Candal/Coelheira. Fase de exploração—Ano 2007	S. Pedro, Casais and Candal/Coelheira	Bats	Not defined (2007)
2012	ProSistemas	Monitorização de quirópteros—Parque Eólico de Serra do Barroso III. Fase de exploração—2º ano	Serra do Barroso III	Bats	2nd year (2011)
2012	PROCESL	Monitorização de quirópteros do Parque Eólico do Bairro (Pias Longas). Relatório 3—2º ano da fase de exploração—2011	Bairro	Bats	2nd year (2011)
2012	LEA	Relatório de Monitorização da Atividade e Mortalidade de Quirópteros no Parque Eólico de Mafômedes. Relatório Anual (Ano 4. Fase de exploração) elaborado para a EnergieKontor	Mafômedes	Bats	4th year (2011)
2012	Ecosativa	Plano de Monitorização de avifauna e quirópteros do Parque Eólico de Fonte da Mesa II. Fase de exploração. 2º Relatório Anual para a empresa Eólica do Alto Douro	Fonte da Mesa II	Birds and bats	2nd year (2011)

(continued)

Table 7.5 (continued)

Publication year	Authors	Title	Wind farm	Faunal group	Monitoring programme phase
2012	Ecosativa	Monitorização de avifauna e quirópteros do Parque Eólico de Cadafaz II. Relatório Final para a empresa ENEOP2 (Outubro 2008—Março 2012)	Cadafaz II	Birds and bats	Final monitoring report
2012	LEA	Relatório de Monitorização da Atividade e Mortalidade de Quirópteros no Parque Eólico do Alto do Marco. Relatório Anual (Ano 1—Fase de exploração) elaborado para a empresa Parque Eólico de Gevancas	Alto do Marco	Bats	1st year (2011)
2012	Plecotus	Parques Eólicos de Bravo e de Mougueiras. Monitorização de quirópteros. Relatório 2—Anos 2009, 2010 e 2011 (relatório final)	Bravo and Mougueiras	Bats	Final monitoring report
2012	LEA	Monitorização da Atividade e Mortalidade de Quirópteros no Parque Eólico de Negrelo e Guilhado. Relatório Anual do 4º ano da fase de exploração—2012	Negrelo e Guilhado	Bats	4th year (2012)
2012	Bio3	Monitorização das Comunidades de Aves e Quirópteros do Parque Eólico de Carreço-Outeiro. Relatório III (Fase III - exploração)	Carreço-Outeiro	Birds and bats	3rd year (2011)
2012	Bio3	Monitorização da Comunidade de Quirópteros do Parque Eólico de Chão-Falcão II. Relatório 3 (Fase de exploração)	Chão Falcão II	Bats	3rd year (2011)

(continued)

Table 7.5 (continued)

Publication year	Authors	Title	Wind farm	Faunal group	Monitoring programme phase
2012	Bio3	Monitorização da Comunidade de Quirópteros do Parque Eólico de Chão-Falcão III. Relatório II (Fase de exploração)	Chão Falcão III	Bats	2nd year (2011)
2012	Bio3	Monitorização das Comunidades de Aves e Quirópteros do Parque Eólico da Nave. Relatório II (Fase III—1º ano de exploração)	Nave	Birds and bats	1st Year (2011)
2012	Ecosativa	Monitorização de avifauna, quirópteros, flora e vegetação do Parque Eólico do Toutiço e respectivas linhas eléctricas e Parque Eólico de Vale Grande. Fase de exploração. Relatório de Progresso	Toutiço	Birds and bats	Not defined
2012	Plecotus	Parque Eólico de Mosqueiros II. Monitorização de Quirópteros. Relatório 3—Ano 2010 e 2011 (relatório final)	Mosqueiros II	Bats	Final monitoring report
2012	Naturiberica	Monitorização Ambiental do Parque Eólico de Alto da Folgorosa—Fase de exploração. Avifauna, Morcegos, Flora e Vegetação. Adenda ao relatório anual 2011	Alto da Folgorosa	Birds and bats	Not defined
2012	Plecotus	Parques Eólicos da Serra da Freita (Freita I e Freita II). Monitorização de Quirópteros: Relatório 4—Ano 2008 e 2009 (relatório final)	Serra da Freita	Bats	Final monitoring report
2012	Bio3	Monitorização da Comunidade de Aves do Parque Eólico de Chão-Falcão III. Relatório IV (Fase de exploração)	Chão Falcão III	Birds	4th year (2011)

(continued)

Table 7.5 (continued)

Publication year	Authors	Title	Wind farm	Faunal group	Monitoring programme phase
2013	ProSistemas	Parque Eólico de Chão Falcão. Monitorização de quirópteros. Relatório nº 5—2008	Chão Falcão	Bats	5th year (2008)
2013	Plecotus (Alves P, Silva B, Barreiro S)	Parque Eólico da Lousã I. Monitorização de Quirópteros. Relatório 2—Ano 2009 (relatório final)	Lousã I	Bats	Final monitoring report
2013	Plecotus (Alves P, Silva B, Barreiro S)	Parque Eólico de Chão Falcão I. Monitorização de quirópteros. Relatório 6—Ano 2009 (relatório final)	Chão Falcão	Bats	Final monitoring report
2013	Strix	Monitorização de Quirópteros na área do Parque Eólico do Barão de São João. Relatório Ano 6—2012. Revisão 1	Barão de São João	Bats	6th year (2012)
2013	Noctula	Relatório de monitorização dos sistemas ecológicos na área do Parque Eólico de Coentral-Safra (Fase de exploração (ano II)—2012/2013)	Coentral-Safra	Birds and bats	2nd year (2012/2013)
2013	PROCESL	Parque Eólico da Raia (Sub-Parques de Benespera, Pousafoles e S. Cornélio)—Monitorização de Quirópteros. Relatório 4 (2º ano da fase de exploração—2012)	Raia	Bats	2nd year (2012)
2013	PROCESL	Monitorização de Quirópteros do Parque Eólico de Alvaiázere. Relatório Anual (Ano 2 da Fase de Exploração—2012)	Alvaiázere	Bats	2nd year (2012)
2013	Bio3	Monitorização da Comunidade de Quirópteros do Parque Eólico de Bornes. Relatório Final (3º ano de exploração)	Bornes	Bats	3rd Year (2012)

(continued)

Table 7.5 (continued)

Publication year	Authors	Title	Wind farm	Faunal group	Monitoring programme phase
2013	Noctula	Relatório de Monitorização dos Sistemas Ecológicos na área do Parque Eólico de Testos II (fase de exploração (ano I)—2011/2012)	Testos II	Birds and bats	1st year (2011/2012)
2013	Ecosativa	Plano de monitorização de Avifauna e Quirópteros do Parque Eólico de Alto Arganil. Fase de exploração	Alto Arganil	Birds and bats	Not defined
2013	LEA	Relatório de Monitorização da Atividade e Mortalidade de Quirópteros no Parque Eólico do Alto do Marco. Relatório Final (Ano 2—Fase de exploração) elaborado para a empresa Parque Eólico de Gevancas	Alto do Marco	Bats	2nd year (2012)
2013	PROCESL	Parque Eólico da Lourinhã II—Monitorização de Quirópteros. Relatório 2—2011/2012 (1° Ano da Fase de Exploração)	Lourinhã II	Bats	1st year (2011/2012)
2013	Bio3	Monitorização das Comunidades de Aves e Quirópteros do Parque Eólico da Nave. Relatório III—Relatório Final (Fase III—2° ano de exploração)	Nave	Birds and bats	2nd year (2012)
2013	Bio3	Monitorização da Comunidade de Quirópteros do Parque Eólico de Chão-Falcao II. Relatório 4 (Fase de exploração)	Chão Falcão II	Bats	4th year (2012)

(continued)

Table 7.5 (continued)

Publication year	Authors	Title	Wind farm	Faunal group	Monitoring programme phase
2013	Noctula	Monitorização da atividade e mortalidade de aves e quirópteros na área do parque eólico da Tocha (fase de exploração (ano I)—2012/2013)	Tocha	Birds and bats	1st year (2012/2013)
2013	Noctula	Relatório intercalar de monitorização de avifauna e quirópteros do Parque Eólico ENERFER I (Fase de exploração—Ano I—Julho 2013)	ENERFER I	Birds and bats	1st year (2013)
2013	Noctula	Relatório intercalar de monitorização de avifauna e quirópteros do Parque Eólico ENERFER I (Fase de exploração—Ano I—Setembro 2013)	ENERFER I	Birds and bats	1st year (2013)
2013	TPF Planege	Parque Eólico de Gevancas II. Monitorização de quirópteros. Fase de exploração (Fase II)—1° ano. Relatório n° 4	Gevancas II	Bats	1st year (2012)
2013	Noctula	Relatório de monitorização dos sistemas ecológicos na área do Parque Eólico de São Macário II (Fase de exploração (Ano I)—2011/2012)	São Macário II	Birds and bats	1st year (2011/2012)
2013	Noctula	Relatório de monitorização da mortalidade de avifauna e de quirópteros na área do Parque Eólico da Serra d'El Rei (Fase de exploração—2012/2013)	Serra d'El Rei	Birds and bats	Not defined
2013	Ecosativa	Plano de Monitorização de avifauna e quirópteros do Parque Eólico de Alto da Coutada. Fase de exploração. Relatório Final	Alto da Coutada	Birds and bats	Not defined

(continued)

Table 7.5 (continued)

Publication year	Authors	Title	Wind farm	Faunal group	Monitoring programme phase
2013	Ecosativa	Plano de Monitorização de avifauna e quirópteros. Ampliação do Parque Eólico do Açor. Fase de exploração	Açor	Birds and bats	Not defined
2013	Bio3	Monitorização da comunidade de aves no Parque Eólico da Lousã II. Relatório V (Fase de Exploração—Ano 2011–2012). Relatório final	Lousã II	Birds	5th year (2011/2012)
2013	Bio3	Monitorização da comunidade de aves no Parque Eólico de Bornes. Relatório Final (Fase de Exploração)	Bornes	Birds	Final monitoring report
2014	PROCESL	Parque Eólico da Raia (Sub-Parques de Benespera, Pousafoles e S. Cornélio)—Monitorização de Quirópteros. Relatório 5 (3º ano da fase de exploração—2013)	Raia	Bats	3rd year (2013)
2014	Noctula	Relatório final de monitorização dos sistemas ecológicos. Parque Eólico de Coentral-Safra (Fase de exploração (Ano III)—2012/2013)	Coentral-Safra	Birds and bats	3rd year (2012/2013)
2014	Ecosativa	Monitorização de avifauna, quirópteros, flora e vegetação do Parque Eólico do Toutiço e respectivas linhas eléctricas e Parque Eólico de Vale Grande. Fase de exploração. Relatório final	Toutiço	Birds and bats	Final monitoring report
2014	PROCESL	Monitorização de quirópteros do Parque Eólico de Alvaiázere. Relatório anual (Ano 3 da fase de exploração—2013)	Alvaiázere	Bats	3rd year (2013)

(continued)

Table 7.5 (continued)

Publication year	Authors	Title	Wind farm	Faunal group	Monitoring programme phase
2014	Strix	Monitorização de Quirópteros na área do Parque Eólico do Barão de São João. Relatório Ano 7—2013. Revisão 1	Barão de São João	Bats	7th year (2013)
2014	TPF Planege	Parque Eólico de Falperra-Rechãzinha. Monitorização de quirópteros. Fase de exploração (Fase II)—1º ano. Relatório nº 2	Falperra-Rechãzinha	Bats	1st year (2013)
2014	Noctula	Monitorização da atividade e mortalidade de aves e quirópteros na área do parque eólico da Tocha (fase de exploração (Ano II)—2013/2014)	Tocha	Birds and bats	2nd year (2013/2014)
2014	Noctula	Relatório intercalar de monitorização de avifauna e quirópteros do Parque Eólico ENERFER I (Fase de exploração—Ano I—Março 2014)	ENERFER I	Birds and bats	1st year (2014)
2014	Noctula	Relatório intercalar de monitorização de avifauna e quirópteros do Parque Eólico ENERFER I (Fase de exploração - Ano I – Maio 2014)	ENERFER I	Birds and bats	1st year (2014)
2014	Noctula	Relatório intercalar de monitorização de avifauna e quirópteros do Parque Eólico ENERFER I (Fase de exploração—Ano I—Julho 2014)	ENERFER I	Birds and bats	1st year (2014)
2014	Noctula	Relatório intercalar de monitorização de avifauna e quirópteros do Parque Eólico ENERFER I (Fase de exploração—Ano I—Setembro 2014)	ENERFER I	Birds and bats	1st year (2014)

(continued)

Table 7.5 (continued)

Publication year	Authors	Title	Wind farm	Faunal group	Monitoring programme phase
2014	Bio3	Monitorização das comunidades de aves e quirópteros no Parque Eólico do Malhanito. Relatório 3 (Fase de exploração—2012/2013)	Malhanito	Birds and bats	3rd year (2012/2013)
2014	Bio3	Monitorização das comunidades de aves e quirópteros no Parque Eólico da Terra Fria—Relatório final de Montalegre	Terra Fria	Birds and bats	Final monitoring report
2014	Noctula	Relatório de monitorização dos sistemas ecológicos na área do Parque Eólico de São Macário II (Fase de exploração (Ano III)—Novembro de 2014	São Macário II	Birds and bats	3rd year (2014)
2014	Noctula	Relatório de monitorização da mortalidade de avifauna e de quirópteros na área do Parque Eólico da Serra d'El Rei (Fase de exploração—2013/2014)	Serra d'El Rei	Birds and bats	Not defined
2014	Noctula	Relatório de monitorização dos sistemas ecológicos na área do Parque Eólico de Testos II (Fase de exploração (Ano II)—2012/2013)	Testos II	Birds and bats	2nd year (2012/2013)
2014	Noctula	Relatório de monitorização dos sistemas ecológicos na área do Parque Eólico de São Macário II (Fase de exploração - Ano II)—2012/2013	São Macário II	Birds and bats	2nd year (2012/2013)
2014	Ecosativa	Plano de Monitorização de avifauna e quirópteros. Ampliação do Parque Eólico do Açor. Fase de exploração	Açor	Birds and bats	2nd year (2013)

(continued)

Table 7.5 (continued)

Publication year	Authors	Title	Wind farm	Faunal group	Monitoring programme phase
2014	Bio3	Monitorização das comunidades de aves no Parque Eólico de Chão Falcão III. Relatório final (Fase de Exploração)	Chão Falcão III	Birds	Operational phase—final report
2014	Bio3	Monitorização da comunidade de quirópteros no Parque Eólico de Meroicinha II. Relatório III (Fase de Exploração)	Meroicinha II	Bats	3rd year (2013)
2015	LEA	Monitorização da Atividade e Mortalidade de Quirópteros no Parque Eólico de Negrelo e Guilhado. Relatório do 4º ano (2012) e final da fase de exploração. Revisão da sua versão original de 2012	Negrelo e Guilhado	Bats	Final monitoring report
2015	LEA	Relatório de Monitorização da Atividade e Mortalidade de Quirópteros no Parque Eólico do Alto do Marco. Relatório Final (Ano 3—Fase de exploração—Ano adicional) elaborado para a empresa Parque Eólico de Gevancas	Alto do Marco	Bats	3rd year (2014)
2015	PROCESL	Monitorização de quirópteros do Parque Eólico de Alvaiázere. Relatório anual 2014 (Ano 4 da fase de exploração)	Alvaiázere	Bats	4th year (2014)
2015	Strix	Monitorização de Quirópteros na área do Parque Eólico do Barão de São João. Relatório Ano 8—2014. Revisão 1	Barão de São João	Bats	8th year (2014)

(continued)

Table 7.5 (continued)

Publication year	Authors	Title	Wind farm	Faunal group	Monitoring programme phase
2015	TPF Planege	Parque Eólico de Falperra-Rechãzinha. Monitorização de quirópteros. Fase de exploração (Fase III)—2º ano. Relatório nº 3	Falperra-Rechãzinha	Bats	3rd year (2014)
2015	PROCESL	Parque Eólico da Raia—reforço de potência. Monitorização de Quirópteros. Relatório 6 (4º ano da fase de exploração—2014)	Raia	Bats	4th year (2014)
2015	Noctula	Relatório de monitorização da atividade e mortalidade de quirópteros na área do Parque Eólico do Sobrado (Fase de exploração – Ano de 2014)—Janeiro de 2015	Sobrado	Bats	Not defined
2015	Noctula	Relatório intercalar de monitorização de avifauna e quirópteros do Parque Eólico ENERFER I (Fase de exploração—Ano II—Março 2015)	ENERFER I	Birds and bats	2nd year (2015)
2015	Noctula	Relatório intercalar de monitorização de avifauna e quirópteros do Parque Eólico ENERFER I (Fase de exploração—Ano II—Maio 2015)	ENERFER I	Birds and bats	2nd year (2015)
2015	Noctula	Relatório intercalar de monitorização de avifauna e quirópteros do Parque Eólico ENERFER I (Fase de exploração—Ano II—Julho 2015)	ENERFER I	Birds and bats	2nd year (2015)

(continued)

Table 7.5 (continued)

Publication year	Authors	Title	Wind farm	Faunal group	Monitoring programme phase
2015	Noctula	Relatório intercalar de monitorização de avifauna e quirópteros do Parque Eólico ENERFER I (Fase de exploração - Ano II – Setembro 2015)	ENERFER I	Birds and bats	2nd year (2015)
2015	Bio3	Monitorização das comunidades de aves e quirópteros no Parque Eólico do Malhanito	Malhanito	Birds and bats	4th year (2013/2014)
2015	Noctula	Relatório Final de Monitorização da Atividade e Mortalidade de Aves e Quirópteros do Parque Eólico da Tocha (Fase de Exploração—Ano III)	Tocha	Birds and bats	3rd year (2014)
2015	Noctula	Relatório de Monitorização da Mortalidade de Avifauna e Quirópteros no âmbito da extensão das pás dos aerogeradores do Parque Eólico da Lameira (Situação de Referência)	Lameira	Birds and bats	Pre-construction
2015	Bio3	Monitorização das comunidades de aves e quirópteros no Parque Eólico de Mosqueiros II. Relatório 7 (4º ano de monitorização da Fase de Exploração—Ano 2014)	Mosqueiros II	Birds and bats	4th year (2014)

References

Amaral, S. M. S. (2009). Análise comparativa da avaliação de impacto ambiental de parques eólicos em Portugal [Comparative analysis of wind farms environmental impact assessment in Portugal]. *Dissertation for MSc in Ecology and Environmental Management*. Faculty of Science of Lisbon University. Lisbon, Portugal

APA. (2010). Guia Para Avaliação De Impacte Ambiental em Parques Eólicos [Methodological Guide to the Environmental Impact Assessment for Wind Farms]. Agência Portuguesa do Ambiente [Portuguese Environmental Agency]. Retrieved from http://www.apambiente.pt/ (in Portuguese).

Bastos, R., Pinhanços, A., Santos, M., Fernandes, R. F., Vicente, J. R., Morinha, F., et al. (2015). Evaluating the regional cumulative impact of wind farms on birds: How can spatially explicit dynamic modelling improve impact assessments and monitoring? *Journal of Applied Ecology, 53,* 1330–1340.

Bastos, R., Santos, M., & Cabral, J. A. (2013). A new stochastic dynamic tool to improve the accuracy of mortality estimates for bats killed at wind farms. *Ecological Indicators, 34,* 428–440.

Bernardino, J., Bispo, R., Costa, H., & Mascarenhas, M. (2013). Estimating bird and bat fatality at wind farms: A practical overview of estimators, their assumptions and limitations. *New Zealand Journal of Zoology, 40,* 63–74.

Bispo, R., Bernardino, J., Marques, T., & Pestana, D. (2013). Modeling carcass removal time for avian mortality assessment in wind farms using survival analysis. *Environmental and Ecological Statistics, 20,* 147–165.

Carrete, M., Sánchez-Zapata, J. A., Benítez, J. R., Lobón, M., Montoya, F., & Donázar, J. A. (2012). Mortality at wind-farms is positively related to large-scale distribution and aggregation in griffon vultures. *Biological Conservation, 145*(1), 102–108.

Dalthorp, D., Huso, M., Dail, D., & Kenyon, J. (2014). *Evidence of absence software. Data series.* US Geological Survey. Retrieved from https://pubs.er.usgs.gov/publication/ds881

Erickson, W. P., Johnson, G. D., Strickland, M. D., & Kronner, K. (2000). Avian and bat mortality associated with the Vansycle wind project, Umatilla County, Oregon: 1999 study year. Technical report prepared by WEST Inc. for Umatilla County Department of Resource Services and Development, Pendleton, OR (25 pp).

Erickson, R. A., Thogmartin, W. E., & Szymanski, J. A. (2014a). BatTool: An R package with GUI for assessing the effect of white-nose syndrome and other take events on Myotis spp. of bats. *Source Code for Biology and Medicine, 9,* 1–10.

Erickson, W. P., Wolfe, M. M., Bay, K. J., Johnson, D. H., & Gehring, J. L. (2014b). A comprehensive analysis of small-passerine fatalities from collision with turbines at wind energy facilities. *PLoS ONE, 9,* e107491.

Etterson, M. A. (2013). Hidden Markov models for estimating animal mortality from anthropogenic hazards. *Ecological Applications, 23,* 1915–1925.

Horvitz, D. G., & Thompson, D. J. (1952). A generalization of sampling without replacement from a finite universe. *Journal of the American Statistical Association, 47,* 663–685.

Huso, M. P. (2011). An estimator of wildlife fatality from observed carcasses. *Environmetrics, 22,* 318–329.

Huso, M. P., & Dalthorp, D. H. (2014). Accounting for unsearched areas in estimating wind turbine-caused fatality. *Journal of Wildlife Management, 78,* 347–358.

Huso, M. P., Dalthorp, D., Dail, D., & Madsen, L. (2015). Estimating wind-turbine-caused bird and bat fatality when zero carcasses are observed. *Ecological Applications, 25,* 1213–1225.

Huso, M., Dalthorp, D., Miller, T. J., & Bruns, D. (2016). Wind energy development: methods to assess bird and bat fatality rates post-construction. *Human-Wildlife Interactions, 10,* 62–70.

Huso M, Som N, Ladd L (2012) Fatality estimator user's guide: U.S. Geological Survey Data Series 729, 22 p

IFC. (2013). *Cumulative impact assessment and management: Guidance for the private sector in emerging markets. Good practice handbook* (102 pp). Washington, DC.

Jain, A., Kerlinger, P., Curry, R., & Slobodnik, L. (2007). Annual report for the Maple Ridge wind power project: Post-construction bird and bat fatality study—2006. Final report. Syracuse, New York, Curry & Kerlinger, LLC (53 pp).

Kerns, J., Erickson, W. P., & Arnett, E. B. (2005). Bat and bird fatality at wind energy facilities in Pennsylvania and West Virginia. In E. B. Arnett (Ed.), *Relationships between bats and wind turbines in Pennsylvania and West Virginia: an assessment of bat fatality search protocols, patterns of fatality, and behavioural interactions with wind turbines* (pp. 24–95). Final report submitted to the Bats and Wind Energy Cooperative. Austin, TX, Bat Conservation International.

Korner-Nievergelt, F., Behr, O., Brinkmann, R., Etterson, M. A., Huso, M. P., Dalthorp, D., et al. (2015). Mortality estimation from carcass searches using the R-package carcass—A tutorial. *Wildlife Biology, 21,* 30–43.

Korner-Nievergelt, F., Brinkmann, R., Niermann, I., & Behr, O. (2013). Estimating bat and bird mortality occurring at wind energy turbines from covariates and carcass searches using mixture models. *PLoS ONE, 8,* e67997.

Korner-Nievergelt, F., Korner-Nievergelt, P., Behr, O., Niermann, I., Brinkmann, R., & Hellriegel, B. (2011). A new method to determine bird and bat fatality at wind energy turbines from carcass searches. *Wildlife Biology, 17,* 350–363.

Masden, E. A., Fox, A. D., Furness, R. W., Bullman, R., & Haydon, D. T. (2010). Cumulative impact assessments and bird/wind farm interactions: Developing a conceptual framework. *Environmental Impact Assessment Review, 30*(1), 1–7.

Meyer, J. R. (1978). *Effects of transmission lines on bird flight behaviour and collision mortality.* Portland, Oregon, USA: Western Interstate Commission for Higher Education-Bonneville Power Administration.

Orloff, S., & Flanerry, A. (1992). Wind turbine effects on avian activity, habitat use, and mortality in Altamont Pass and Solano County Wind Resource Areas 1989-1991. Final report. Tiburon, CA, Biosystems Analysis Inc., California Energy Commission (164 pp).

Péron, G., & Hines, J. E. (2014). fatalityCMR—Capture-recapture software to correct raw counts of wildlife fatalities using trial experiments for carcass detection probability and persistence time: U.S. Geological Survey Techniques and Methods 7–C11 (14 pp).

Péron, G., Hines, J. E., Nichols, J. D., Kendall, W. L., Peters, K. A., & Mizrahi, D. S. (2013). Estimation of bird and bat mortality at wind-power farms with superpopulation models. *Journal of Applied Ecology, 50,* 902–911.

Poot, M., Van Horssen, P., Collier, M., Lensink, R., & Dirksen, S. (2011). Effect studies offshore wind Egmond aan Zee: Cumulative effects on seabirds. A modelling approach to estimate effects on population levels in seabirds. Bureau Waardenburg bv. Culemborg, The Netherlands.

Ramos Pereira, M. J., Salgueiro, P., Rodrigues, L., Coelho, M. M., & Palmeirim, J. M. (2009). Population structure of a cave-dwelling bat, *Miniopterus schreibersii*: Does it reflect history and social organization? *Journal of Heredity, 100,* 533–544.

R Core Team. (2016). R: A language and environment for statistical computing. R Foundation for Statistical Computing, Vienna, Austria. Retrieved from https://www.R-project.org/

Rodrigues, L., Bach, M. J., Dubourg-Savage, B., Karapandza, D., Kovac, T., Kervyn, J., et al. (2015). Guidelines for consideration of bats in wind farm projects—Revision 2014. EUROBATS Publication Series No. 6 (English Version). UNEP/EUROBATS Secretariat, Bonn, Germany (133 pp).

Rodrigues, L., & Palmeirim, J. M. (2008). Migratory behaviour of the Schreiber's bat: When, where and why do cave bats migrate in a Mediterranean region? *Journal of Zoology, 274,* 116–125.

Roscioni, F., Russo, D., Di Febbraro, M., Frate, L., Carranza, M. L., & Loy, A. (2013). Regional-scale modelling of the cumulative impact of wind farms on bats. *Biodiversity and Conservation, 22*(8):1821–1835.

Santos, H., Rodrigues, L., Jones, G., & Rebelo, H. (2013). Using species distribution modelling to predict bat fatality risk at wind farms. *Biological Conservation, 157*, 178–186.

Shoenfeld, P. (2004). Suggestions regarding avian mortality extrapolation. Technical memo provided to FPL Energy. Davis, WV, West Virginia Highlands Conservancy (6 pp).

Smallwood, K. S., Bell, D. A., Snyder, S. A., & DiDonato, J. E. (2010). Novel Scavenger removal trials increase wind turbine-caused avian fatality estimates. *Journal of Wildlife Management, 74,* 1089–1097.

Stevens, B. S., & Dennis, B. (2013). Wildlife mortality from infrastructure collisions: Statistical modeling of count data from carcass surveys. *Ecology, 94,* 2087–2096.

Strickland, M. D., Arnett, E. B., Erickson, W. P., Johnson, D. H., Johnson, G. D., Morrison, M. L., et al. (2011). Comprehensive guide to studying wind energy/wildlife interactions. Prepared for the National Wind Coordinating Collaborative, Washington, DC (281 pp).

Voigt, C. C., Lehnert, L. S., Petersons, G., Adorf, F., & Bach, L. (2015). Wildlife and renewable energy: German politics cross migratory bats. *European Journal of Wildlife Research, 61,* 213–219.

Wolpert, R. L. (2015). ACME: A partially periodic estimator of Avian & Chiropteran mortality at wind turbines. 1–37. Retrieved from http://arxiv.org/abs/1507.0749

Zimmerling, J. R., & Francis, C. M. (2016). Bat mortality due to wind turbines in Canada. *The Journal of Wildlife Management, 80,* 1360–1369.

Chapter 8
How to Design an Adaptive Management Approach?

Helena Coelho, Silvia Mesquita and Miguel Mascarenhas

8.1 What Is Adaptive Management?

Adaptive management was first proposed in the early 1900s and its earliest relation to the natural resources management is found in publications of the 70s and 80s (Stankey et al. 2005; van der Brugge and van Raak 2007).

Currently, there are several studies that provide information on the use of adaptive management approaches to resource management (Williams and Brown 2016). Adaptive management is widely considered to be one of the best available solutions for managing biological systems, although in the presence of uncertainty (Westgate et al. 2013). Adaptive management practices are also included in the Performance Standards on Environmental and Social Sustainability (e.g., Performance Standard 6) by International Finance Corporation for managing environmental and social risks and impacts (IFC 2012) or by IUCN (2016) in its Standard on Biodiversity Conservation and Sustainable Use of Natural Resources.

The increase of references and recommendations for the use of the adaptive management approach is directly related to the current challenges faced by natural resource managers, such as the increasing need to conserve and promote biologic diversity, react to climate change or provide the best solutions to restoration.

The adaptive management strategy applied to biodiversity might be thought of as a systematic process of improving management through continuous learning. Adaptive management emphasises the importance of accounting for the future

H. Coelho (✉) · S. Mesquita · M. Mascarenhas
Bioinsight, Odivelas, Portugal
e-mail: helena.c@bioinsight.pt

S. Mesquita
e-mail: silvia.m@bioinsight.pt

M. Mascarenhas
e-mail: miguel.m@bioinsight.pt

M. Mascarenhas et al. (eds.), *Biodiversity and Wind Farms in Portugal*,
https://doi.org/10.1007/978-3-319-60351-3_8

consequences of present actions, by addressing the uncertainty and looking for ways to reduce this uncertainty (Williams and Johnson 2013; Williams and Brown 2016). Considering the management of natural resources, adaptive management might be presented as 'learning by doing' and adapting activities based on new information learned (Walters and Holling 1990; Williams and Brown 2012, 2016). It also can be seen as a "natural, intuitive and potentially effective way to make decisions" (William 2014).

Adaptive management should be based on the recognition that resource systems, especially ecosystem-based systems, are only partially understood. Considering this, there will be an increased value in tracking its conditions and using the results while learning or understanding improves (Williams and Brown 2012). Adaptive management might be extremely useful where natural resources are responsive to management but uncertainty still exist regarding the impact of management actions (Williams and Brown 2012).

According to the international guidelines (Williams et al. 2007; Williams and Brown 2012), adaptive management is described with a **set-up phase**, corresponding to planning, during which some essential components will be put in place, and an **iterative phase**, which proceeds to an on-going cycle of learning about the structure and function of the system. The adaptive management set-up includes stakeholder involvement, objectives, management alternatives, predictive models and monitoring protocols. Steps for the iterative phase include decision making, follow-up monitoring, assessment and learning, as well as feedback.

The concept of adaptive management is relatively new in the management of biodiversity **applied to wind energy projects**. The first studies in international guidelines to the application of this approach in wind energy were carried out by the European Union (European Commission 2011a) and the USA (Strickland et al. 2011; USFWS 2012). These guidelines clearly recommend wind energy projects to be developed in an adaptive management context, reinforcing also the need for adaptive decision making and for a proactive way to answer challenges.

In **Portugal**, the concept of adaptive management has been applied in a few wind farms, even though they were not initially planned under adaptive management assumptions (see information on the Portuguese Case Study—Sect. 8.4). This mainly results from an active decision-making strategy, a consequence of the collaboration of several stakeholders (e.g., ecological specialists, developers, regulators/authorities).

However, adaptive management it is not formally adopted in the Environmental Impact Assessment (EIA) process and legislation of Portugal (*see information* on Portuguese Legislation on Chap. 3). However, technical requirements of the monitoring report (defined by Portaria no. 395/2015 of November, 4th) allows the (re)evaluation of impacts and effectiveness of measures implemented considering the results of the monitoring plan. It also grants an opportunity to adjust the monitoring plan and the definition of additional mitigation measures.

Despite the fact that several studies highlighted the need for and the advantages of the use of this adaptive management approach to solve resource problems, and the fact that this approach has now begun to be followed in management or conservation plans of wind energy projects, adaptive management faces significant challenges and the record of success for its implementation remains relatively limited (William 2016, 2014 and Westgate et al 2013). In addition, frequently, the knowledge learned is not included in the currently on-going projects but rather in the subsequent projects (Bulling and Koppel 2016).

8.2 Adaptive Management of Wind Farm Impacts on Biodiversity

Wind energy is virtually inexhaustible and relatively free of carbon emissions and contributes positively to the efforts to counteract climate change caused by increasing human energy needs. However, this source of energy has some negative impacts on biodiversity, especially on birds (Drewitt and Langston 2008; Martínez et al 2010; Marques et al. 2015) and bats (Barclay et al. 2007; Arnett et al. 2011, 2015). The most significant impacts include the mortality due to collision, barrier effects and disturbance (Drewitt and Langston 2006; Baerwald et al. 2008; Marques et al. 2015; see Pereira et al. Chap. 2).

Facing the need to overcome these impacts, biological systems are often managed so that impacts can be mitigated and, ideally, the state of the ecosystem can be maintained (no net loss) or improved (net gain) (BBOP 2012). However, many potential management actions can be done and uncertainty about which one(s) will be most effective is common in all projects. Therefore, managers often have to make a trade-off between actions that are expected to maintain or improve the system state, and actions designed to improve knowledge (Westgate 2013).

There are six major contributors to an adaptive management approach: impact-related aspects, including wind farm specificities, site characteristics and species factors; and processes, including impact assessment, mitigation hierarchy and monitoring (Santos et al. 2015).

As Bennet et al. (2016) recently highlighted, significant changes at an institutional level are also required to achieve goals with adaptive management. Both cultural and technical changes, which also involve changing attitudes, are needed to understand the risk-based decision making in order to address the challenges presented by the uncertainties of the impacts. This decision-making approach also requires a collaborative and inclusive engagement of developers, regulators and other interested stakeholders.

Applying adaptive management to a given wind power project allows a conscious combination of actions that simultaneously promote ecosystems maintenance or improvement while increasing knowledge in what works best in a specific scenario. To do so, two major components have to be extensively explored: mitigation hierarchies and monitoring programmes.

8.2.1 Mitigation Hierarchy

Mitigation typically involves a sequence of well-known processes, activities or actions to avoid, reduce or offset significant negative impacts caused by a development project, such as wind farms (Marshall 2001; Arnett and May 2016).

Although there are various definitions for mitigation, one of the most accepted among specialists encompasses four sequential levels of action, known as 'mitigation hierarchy': (i) prevent/avoid; (ii) minimise (iii) rehabilitate/restore; and (iv) compensate/offset (BBOP 2013; IFC 2012) (Fig. 8.1).

According to this definition, efforts to avoid impacts should always be prioritised. Avoidance measures in wind farms frequently include macro or micrositing options, such as site-specific project planning to exclude wind turbines from protected or sensitive areas (e.g., bat roosts, bird migratory routes, nesting places or protected habitats) (Gartman et al. 2016b). The persistent negative effects should be minimised as far as possible and actions should be taken to rehabilitate or restore degraded biodiversity. This might be achieved through the implementation of different measures, such as deterrence systems (e.g., acoustic devices, electromagnetic fields), shutdown on demand or curtailment (Gartman et al. 2016b; GPWind 2013; Marques et al. 2014). In parallel, measures regarding restoration or rehabilitation, such as the control of exotic plants or planting native trees or seeds to recover natural habitats (Paula et al. 2015). Lastly, the acceptable residual impacts that cannot be avoided or minimised should be the subject of compensation/offset measures. Examples of compensation/offsets measures include habitat enhancement through prey population management for impacts on top predators (Santos et al. 2012) or enhancement of riparian galleries for bat populations (Peste et al. 2015).

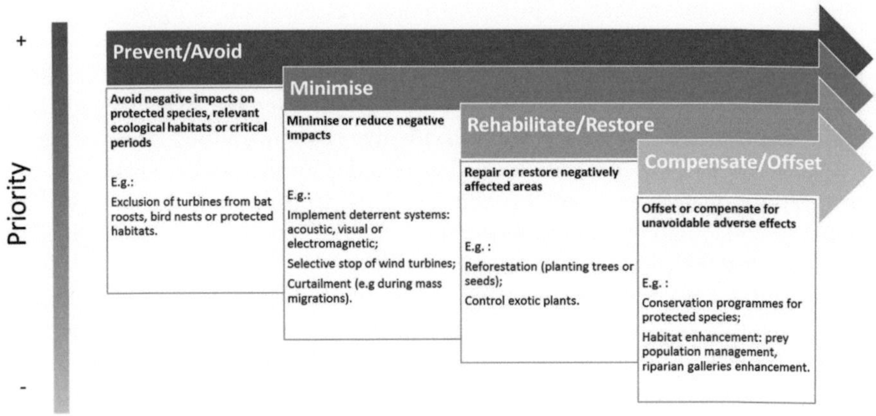

Fig. 8.1 Mitigation hierarchy adapted to wind energy projects and examples of measures reflecting good practices per level: avoidance, minimisation, rehabilitation/restoration and compensation/offset

Mitigation measures in a context of decision-making resulting from collaborative approach between regulators, developers and other stakeholders have been implemented in **Portugal** over the last 10 years (Table 8.1). It is also important to highlight the fact that some of these practices were sustained in a dynamic approach, including monitoring results from post-construction or/and a carefully scoping phase.

Mitigation hierarchy should be considered from the initial phase of the project design, i.e. for the identification of Environmental Constraints or EIAs (EC 2011a, b), to ensure that the project is located in areas with a lower associated impact and the implementation of any compensatory measures begins before the construction phase. To ensure greater efficiency, the mitigation actions should be based on practical application of scientific understanding (Arnett and May 2016). However, regardless of the efficiency of each known type of measure, the mitigation strategy should consider the ecological, social and economic context of the region where the infrastructure will be located (BBOP 2012). Monitoring and evaluation of the effectiveness and impacts of mitigation measures is a crucial step for a powerful adaptive management and will be further discussed in Sect. 8.2.2.

As mentioned earlier, the hierarchy of mitigation must be considered from the early stages of the project so that the countervailing measures can be planned in advance and can be sustainable from an ecological, social, economic and legal point of view. One of the most challenging and important tasks in this type of project is stakeholders' (e.g., ecological specialists, developers, authorities) involvement during the various stages of the mitigation hierarchy (BBOP 2012). Conservation might be performed with communities involvement as social concerns increase the risk of failure and reduce the probability to achieve targets and long-term results (Enevoldsen and Sovacool 2016). Consequently, for each project it is essential to understand not only the ecological dimension, but also the societal context to find 'common interests' that can be used to promote the involvement of stakeholders and increase acceptance (Enevoldsen and Sovacool 2016).

8.2.2 Monitoring

Monitoring programmes are a critical step towards a successful adaptive management approach since they improve knowledge that is fundamental to the decision-making process. Monitoring programmes allow us to (i) assess the real impacts of the wind farm and compare them with the impacts that were foreseen before construction; (ii) understand the interactions between the biological communities (e.g., birds, bats, flora and habitats, and wolves) and project structures (e.g., wind turbines and associated power lines); (iii) evaluate the effectiveness of mitigation measures; and (iv) advise on the direction of further measures (Williams and Brown 2012, 2016; Marques et al. 2015). The success of a monitoring programme relies on a good planning and, subsequently, the ability to adjust said programme, in accordance with its results (Fig. 8.2). Monitoring plans should also be focused and avoid general aims or trying to achieve too much at once; this might

Table 8.1 Example of measures related to mitigation hierarchy (levels: avoidance, minimisation, rehabilitation/restoration and compensation/offset) applied in Portugal

Mitigation level	Target species/group	Examples	References
Avoidance	Fauna, flora & habitats	Adjustments to the wind farm layout are frequent in Portugal during the EIA process to avoid negative impacts on ecological values, such as bird nests, protected habitats etc. This evaluation is supported by the Portuguese environmental legislation through the scoping report or EIA report. An integrated approach allows identification of the potential significant impacts that will result from the implementation of the project and its proposed layout alternatives, supporting the decision regarding the environmental viability	EC (2011), APA (2010)
Minimisation	Migration birds	In the south of Portugal (an important crossing point for soaring birds, as they migrate from Western Europe to Africa), minimisation and real-time monitoring have been tested with positive results. Selective stopping techniques at turbines with higher mortality rates were implemented to reduce the negative impacts on birds. Between 2010 and 2013, despite the fact that >25,000 individual flight movements of soaring birds were recorded in one wind farm per year, zero mortality was observed as a result of mitigation measures applied. It is also important to highlight that these measures have been implemented with a minimal effect on energy production	Birdflife International (2015), GPWind (2013), Tomé et al. (2015, 2012)
	Bats	Restriction of turbine operation has been implemented in Portugal regarding the conservation of bats and to reduce bat fatalities. In the north of Portugal at least one wind farm has implemented, on an experimental base, a minimisation measure to six turbines with a cut-in speed at 3.3 m/s between September 15 and October 15 of 2010. Results showed a potential reduction of 31.4% on bat mortality compared with the absence of curtailment	LEA (2010)

(continued)

Table 8.1 (continued)

Mitigation level	Target species/group	Examples	References
Rehabilitation/restoration	Habitats/invasive plants	Rehabilitation/restoration have been successfully tested in central Portugal. Increasing awareness about the invasive plants has motivated a better address of this problem in the EIA process. Several monitoring and control plans have been required and implemented since 2006. Ten years of control measures for invasive plants on wind farms allows identification of efforts that should be made to implement control plans in an early stage in order to increase their success rates and reduce cost. Collaborative engagement of all stakeholders is also a key factor to success	Passos et al. (2013)
Compensation/offset	Eagles/Kestrels	Several compensation programmes have been successfully implemented in Portuguese wind farms over the last ten years. Compensation measures tested habitat management in order to mitigate negative impacts. Planting native scrub underneath turbines to make areas less attractive to kestrels, or the promotion and recovery of European rabbit populations to promote eagle activity outside of wind farm are current measures applied in Portugal. Creation of supplementary food and water provision in the dry season to increase prey availability is also a measure currently applied	Bioinsight (2016), Bio3 (2012), Cordeiro et al. (2013), Marques et al. (2012), Santos et al. (2012)
	Iberian wolf	Since 2007, as a consequence of the wind energy development, Portugal has established several projects regarding wolf compensation in an integrated way. Compensation was promoted through several aspects of wolf habitat conservation, namely forest management, reduction of the conflict between human and wolf, management of wild prey through re-introduction and implementation of restricted hunting-areas	Brotas et al. (2015), Gartman et al. (2016b)

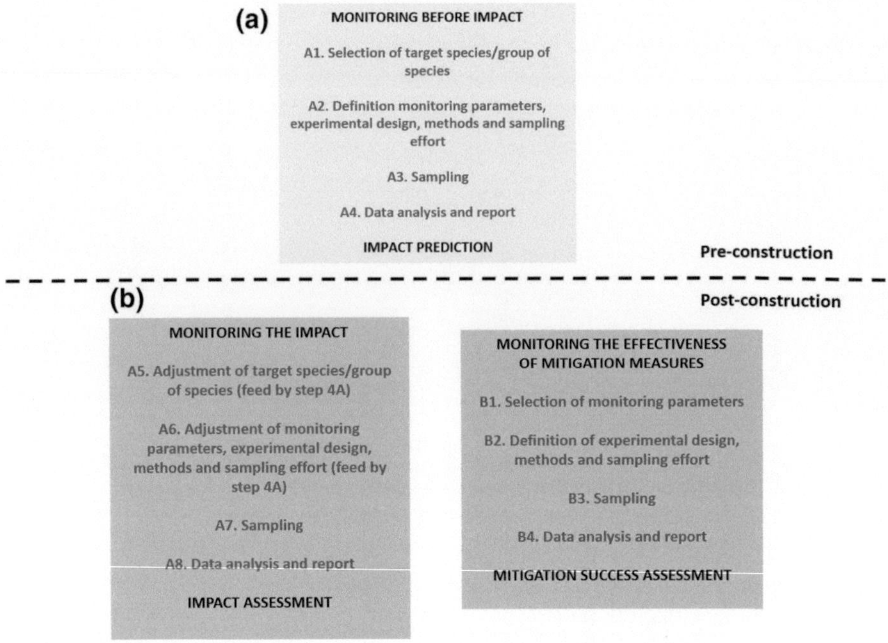

Fig. 8.2 Monitoring of impacts before and after construction (**a**); and the mitigation success assessment (**b**)

be improved by structured and coordinated collaborative approaches (Bennet et al. 2016).

Since adaptive management is a 'learning by doing' process, assessing biological communities' response to impacts and mitigation actions through a sound monitoring programme is a key aspect. Another important aspect is the establishment of a close relationship between monitoring phases (pre- and post-construction) and between monitoring objectives (impacts and mitigation success) (Williams and Brown 2014). An informed decision on what is the best way forward can only happen if all relevant information is presented to all stakeholders.

Bearing in mind that decisions must be sustained by a mitigation hierarchy approach and supported by sound monitoring programmes, it is possible to propose an adaptive management approach for wind energy projects.

8.3 Step-by-Step Adaptive Management Approach

This topic intends to develop and deepen the steps to follow in each phase of an adaptive management approach, with a view to reconciling the wind farms with biodiversity.

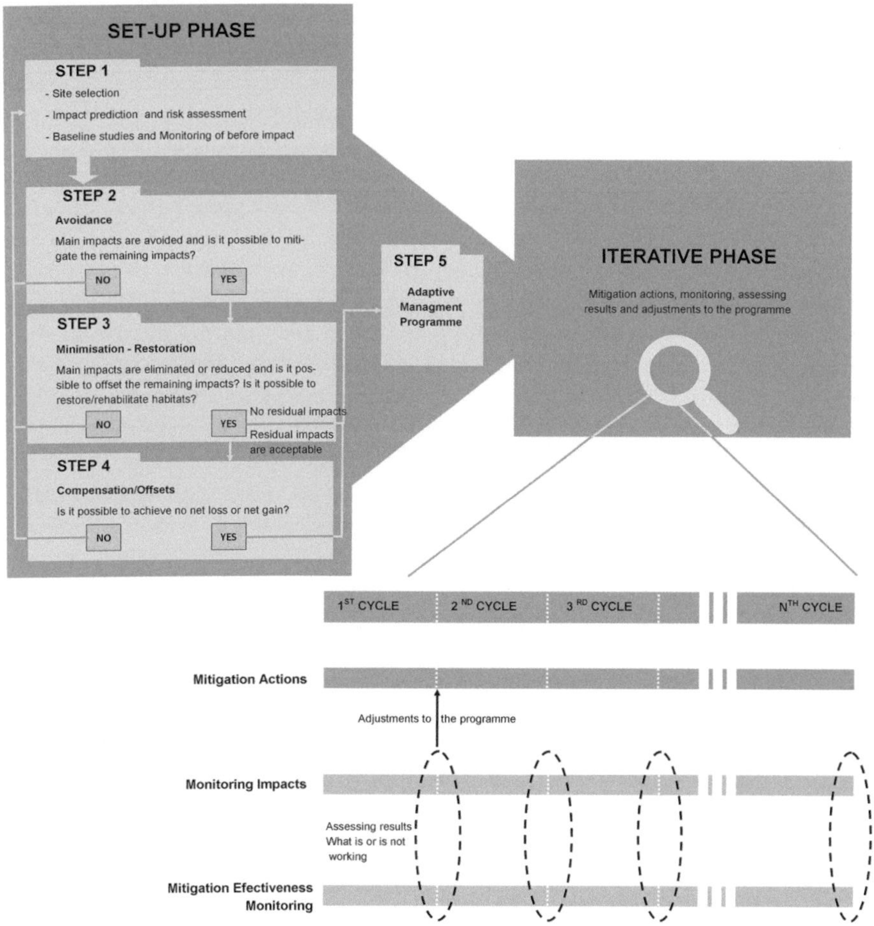

Fig. 8.3 Adaptive management approach scheme for a wind energy project

It is based on international guidelines and integrates both the mitigation hierarchy and the philosophy of adaptive management. Although the proposed scheme in Fig. 8.3 can be adjusted to any project, it is important to note that this approach is based on the premise that each case must be examined individually (Santos et al. 2015). There is no universal solution that can be implemented in all cases due to the high complexity of the parameters to be considered when defining the strategy to be implemented for a given wind farm (Santos et al. 2015).

This approach encompasses two major phases: the set-up phase and the iterative phase (Williams et al. 2009; Williams and Brown 2012, 2014). As mentioned before by Williams and Brown (2012, 2014, 2016), the set-up phase evolves along project planning and is a process whereby different scenarios are assessed. The decisions made during this phase are sustained by baseline information and pre-construction monitoring results and are oriented by the mitigation hierarchy. The iterative phase is a sequential process of action and decision-making, where actions are adjusted to the results of monitoring.

For the **set-up phase**, four to five steps must be completed (Fig. 8.3):

Step 1—This includes site selection (planning and siting), impact prediction and risk assessment. This step is when decisions are orientated towards choosing the best site location in terms of impacts and the risk for biological communities. To this end, it is essential to make an assessment of potential impacts, risks and the careful definition of the goals to be achieved (Gartman et al. 2016a). In most cases, this decision can only be taken after the completion of appropriate baseline studies. This first step should include both bibliographic research and field work (identification of protected areas, the inventory of habitats, flora and fauna). At this stage, it is also important to start the monitoring programme that will systematically collect data about the ecosystem state.

Step 2—This step requires having a site selected and its ecological context assessed (baseline information, impacts and risks) to move into the mitigation hierarchy. At this point, is important to first evaluate avoidance measures (Gartman et al. 2016a). Typical measures of avoidance in wind farms include turbine exclusion from areas intensively used by bats, bird migratory routes, bird of prey nesting places, or protected habitats (Gartman et al. 2016a; Paula et al. 2015). If it is not possible to avoid the major negative impacts, one should return to the beginning of the analysis (step 1) in order to select a new location to place the wind farm. At this early stage, it is essential to ensure the involvement of all stakeholders in the process, including the developer, engineers, experts in terms of wind energy production, as well as experts who will evaluate the impact on the environment and identify necessary mitigation measures (Williams and Brown 2012).

Step 3—This is when main impacts are considered to have been avoided, and it is then possible to proceed to the next step, in which strategies to minimise impacts are defined. At this point, strategies for the restoration/rehabilitation of affected habitats should also be defined. Minimisation measures can include deterrents, selective stopping of turbines, curtailment, increase in blade visibility or even habitat management to decrease activity in the wind farm area (Paula et al. 2015). To enforce the decision-making process, it is recommended that scenarios be established for each minimisation measure that is framed by ecological and environmental thresholds. For instance, when defining curtailment measures, it is an advantage to establish thresholds for bat activity, wind speed and temperature. These thresholds will guide the decision into applying a more or less restrictive curtailment scenario (Marques et al. 2016). Considering that critical decisions are being made, it is essential to ensure that all stakeholders are also involved in this step, working together in defining different scenarios for the minimisation and restoration/rehabilitation, as well as in defining acceptable thresholds (Williams and Brown 2012, 2014). By doing so, rules on how to proceed are jointly defined and are clear for everyone.

After completion of step 3 it is necessary to determine if there are potential residual impacts remaining. If no residual impacts remain, one can proceed to step 5 but if there are still residual impacts that need to be addressed, then their significance determines which step to take. In case those impacts are not acceptable from the conservation point of view, it is necessary to return to step 1. But if the potential residual impacts are acceptable, step 4 must be accomplished.

Step 4—If residual impacts are determined to be acceptable, compensation/offset programmes must be defined. As mentioned in the mitigation hierarchy proposed by BBOP (2102), compensation/offset measures aim to achieve no net loss or even net gain. Therefore, it is crucial to have the best knowledge possible on ecosystem status and what the expected results are for the compensation/offsets. Compensation/offset examples include habitat enhancement through prey population management for impacts on top predators (Santos et al. 2012), electrocution prevention (Cole et al. 2013) or managing coniferous production forests for bat populations (Pereira et al. 2016). Again, all stakeholders should be involved and agree on what the compensation objectives are, as well as the expected results (Williams and Brown 2012).

It is important to stress that each step of the mitigation hierarchy does not replace the previous one and that all of them should be seen in an integrated way.

Bearing all of these actions in mind (from step 1 to step 4), the developer will be able to determine whether its project is or is not economically feasible, even in the worst case scenario and, if it is, we can move along to step 5.

Step 5—This encompasses the preparation of the adaptive management programme. This is a document that will guide not only the execution of the mitigation measures during project construction and operation but will also determine the adequate monitoring programme to assess those same actions. Depending on the laws of each country, the programme could be integrated in the project Environmental Management Plan or could simply replace it when it is not legally recognised (*see more* about Portuguese legislation in Santos el al. Chap. 3). As mentioned before, different stakeholders must agree to the acceptable limits, which will guide the changes that may occur later in the programme. These changes can include additional mitigation measures that may be required once the results of monitoring are considered. It is essential to bear in mind that some steps in this integrated management approach should be actively included and implemented not only in the planning phase, but also during and after construction of the wind farm. This is the core idea of the learn-by-doing approach and this concept needs to be clear for all the stakeholders involved in the decision.

Once step 5 is completed and there is a common understanding among all stakeholders on what should be done during construction and operation, the **iterative phase** begins. The iterative phase is the core of the adaptive management approach—learning by doing. The basis of the approach involves continuous cycles of applying the mitigation hierarchy, monitoring impacts, monitoring the effectiveness of the mitigation measures and periodically assessing all the results, and determining what is working, what needs to be changed and proposing measures necessary to accomplish such a change (Williams and Brown 2012, 2014).

In this context, we highlight that the key to the success of the adaptive management programme is based primarily on the strength of monitoring programmes.

If they are not well designed, the effectiveness will be reduced. It is therefore essential that both the impact and the results of mitigation strategies are properly evaluated, allowing for continuous learning that will, in the long run, keep improving and/or developing new methods of mitigation. It is also imperative that this strategy is implemented through the involvement of all stakeholders, which is the key to the success of the project, ecologically, economically and socially (Williams and Brown 2012, 2014; Santos el al. 2015).

8.4 Case Study—Candeeiros Wind Farm

(*Case Study based on publicly available information of EIA and monitoring reports of Candeeiros wind farm* (*Bioinsight* 2016; *Bio3* 2012, 2007)).

In Portugal, as previously stated, a formal approach to adaptive management is not yet included in the environmental process. Although some instruments consider a linkage between impacts, mitigation and monitoring, the following case study demonstrates a comprehensive approach that goes beyond Portuguese legislation.

Candeeiros wind farm is located in a ridge of a limestone mountain in Central Portugal, inside a Protected Area and a Natura 2000 Site (Parque Natural Serra de Aire e Candeeiros). The area (ca. 70 km^2) comprises calcareous hills with lots of quarries and cliffs. Typical Mediterranean shrubs (e.g., *Quercus coccifera* Linnaeus, *Rosmarinus officinalis* Linnaeus, *Daphne gnidium* Linnaeus) of different heights and densities dominate the vegetation (Paula et al. 2015). The surrounding areas are also composed of eucalyptus tree plantations and small farmland areas.

The wind farm started its operation in 2005 with 26 wind turbines, each with a power of 3 MW. In 2007, 11 turbines were added to the wind farm. During 2015, a repowering project added five more turbines, each with a power of 2 MW. Currently, the overall capacity of this wind farm is 121 MW, with an estimated annual average production of 361 GWh.

In 2003, Candeeiros wind farm went through an EIA process where the main impacts were identified and, in face of uncertainty, a monitoring programme to assess the real impact of the wind farm on birds was established.

In 2005, the wind farm's bird monitoring programme began. During the first cycle of monitoring, which occurred until the end of 2007, a general bird survey was undertaken monitoring the whole bird community. Methodologies included point counts and weekly fatality searches, by humans, around all wind turbines.

The bird community was assessed by two methods: (i) 5 min point counts for the overall bird community and (ii) 1 h point counts dedicated to raptors and other soaring birds. In the case of point counts for raptors and soaring birds, the monitoring objectives were to determine: (i) number of individuals/pairs and (ii) flight behaviour and collision risk.

The bird monitoring programme also included a specific approach to the red-bill chough (*Pyrrhocorax pyrrhocorax*), which is an endangered bird in Portugal. This specific monitoring approach was proposed considering the presence of breeding

Fig. 8.4 Common kestrel (**a**–**c**); and detail of ringing (**b**, **c**)

pairs of chough in the wind farm area, its population decline in the region, in previous years, and the potential impacts of the wind farm on this species. To monitor this species, monthly visits and 1 h point counts within breeding sites were performed.

Fatality searches were carried out on a weekly basis in a 50 m radius around each one of the wind turbines. Searcher efficiency and carcass removal rates were also determined during the monitoring programme so that it was possible to achieve fatality estimates.

Between 2005 and 2007, fatality searches showed a high mortality of just one species, the common kestrel (*Falco tinnunculus*), with seven fatalities found in 3 years. Fatality rates were calculated considering bird monitoring programme results (8–12 individuals of the species in the wind farm area) and the results pointed to an annual average of 30% of the population dying due to collision with the turbines. However, there were no significant changes in the number of common kestrel that used the area during the 3 years.

No other significant impacts were identified for any other species, including the target species that was initially determined: red-bill chough.

Considering these results, in 2008, the monitoring programme was adapted so that it was possible to focus the efforts on a new target species—the common kestrel (Fig. 8.4). However, as a precautionary measure, the chough population also continued to be monitored.

The monitoring for the kestrel population included, at the time, an expansion of the study area, car transects and point counts (1 h each). Breeding sites were also searched and monitored.

This adaptation of the monitoring programme was intended to obtain better population estimates, which would allow better quantification of the impact that the wind farm was having on the population.

In 2011, this method was reinforced with the ringing of the population (Fig. 8.4b, c), allowing the distinction of individuals. During 2011 and 2012, a total of 29 individuals were ringed in the study area. Results of both methods were integrated allowing a population estimation of 9–13 breeding pairs.

Over the years, weekly fatality searches continued and again common kestrels were the most affected species in terms of mortality, with 10 individuals found dead between 2008 and 2012 and a total of 17 individuals found dead since the beginning of the operation phase.

Results during this period (2005–2012) showed fatality estimates of 75–108 dead kestrels (following Korner-Nievergelt et al. 2011 and Huso et al. 2010 formulas, respectively) since the beginning of the wind farm's operation phase. Although there is some uncertainty in population and mortality estimates, and the kestrel is not an endangered species in Portugal, these results could not be ignored and concern was raised regarding the success and viability of the kestrel population at the study area. This concern led to a challenge: the need to find better solutions for the management of the impacts on kestrel populations as a result of wind farm operation.

The challenge was integrated by stakeholders at different levels: experts, developers and authorities and have conducted to the proposal of a site-specific mitigation programme. The proposed programme comprised on-site minimisation and offset/compensation and was developed to reduce kestrel mortality at Candeeiros wind farm (Fig. 8.5).

The knowledge acquired on the common kestrel's population and behaviour in the wind farm and surrounding areas, as a result of the monitoring, was used as crucial information to the development of the programme. This reinforce learning as the key factor in the adaptive management approach.

The mitigation programme was specifically established in order to reduce collision risk and, consequently, the mortality of kestrel species. In order to achieve these aims, habitat management methodologies inside and outside the wind farm area were used.

Two specific objectives were established within the programme: (i) to decrease the level of attractiveness of the areas around turbines and (ii) to increase food availability in areas outside the wind farm and with lower collision risk. Considering this, three specific actions were defined for implementation between 2013 and 2016:

1. Plantation of native scrub species below the turbines to obtain denser vegetation in the area, making it less attractive to hunting kestrels;
2. Opening patches inside the scrub areas to enhance habitat heterogeneity and therefore increase prey density and availability in areas with lower risk of collision;
3. Promotion of extensive grazing by goats, also away from the turbines, to enhance habitat heterogeneity.

These measures were also favourable to the red-bill chough population, since this species selects areas with short and sparse vegetation for feeding.

The implementation of the programme was also sustained by the monitoring of the kestrel population and the fatality searches. Both have continued, in order to evaluate the success of the implemented measures (Fig. 8.5).

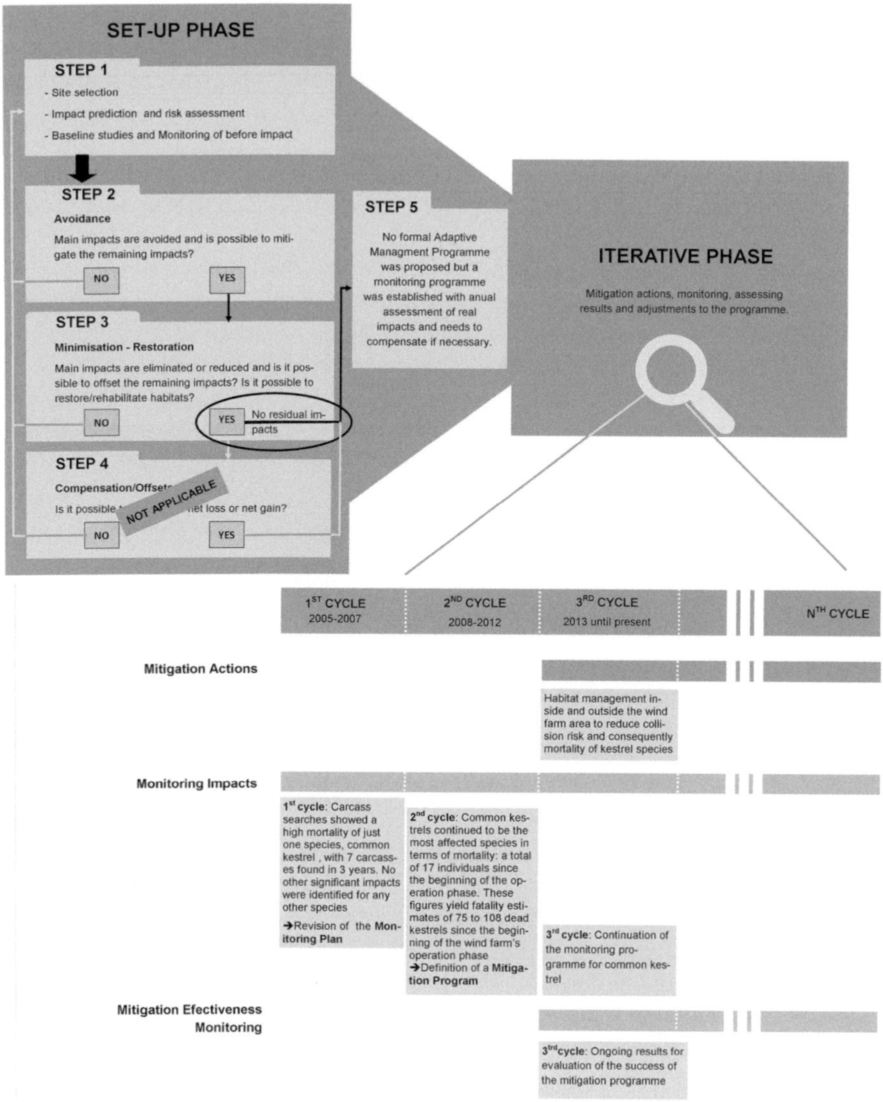

Fig. 8.5 Application of the adaptive management approach for a Portuguese wind farm

Until 2015, more than 18,000 kermes oak (the native scrubs species) were planted, covering approximately 5 ha of the area below turbines. Two management areas were established away from turbines, where the opening of patches and extensive grazing by goats was promoted. All the actions were accomplished with the involvement of local stakeholders (civil groups) and promotion of the local economy.

As for the kestrel population, data between 2013 and 2015 showed only small fluctuations in the breeding population: nine confirmed pairs in 2013, eight in 2014 and ten in 2015. Population estimates, which also include juveniles and adults not breeding in the area, also showed no significant changes in population size.

Regarding kestrel fatalities, during this period (2013–2015), two carcasses were found (one in 2013 and one in 2015) and fatality rates ranged from 8–17% of the population (using Korner-Nievergelt et al. 2011 and Huso et al. 2010 formulas).

On-going results shows that in recent years, kestrel mortality has decreased in the wind farm area; however, it is not yet possible to establish this as a consequence of the success of the measures implement since some of the measures, such as the plantation, are expected to be succeed on a longer timeframe.

8.5 Final Remarks

In recent years, the development of wind farms has faced several challenges in terms of the assessment of their real and perceived impacts on biodiversity.

The development of new approaches that allow the installation of wind farms following good and sustainable environmental practices and promoting a compromise between economic sustainability, biodiversity conservation and social requirements is crucial.

The adaptive management approach has allowed development of a way to reconcile wind energy development with the maintenance of natural values, integrating both the principles of the mitigation hierarchy and the traditional adaptive management concept.

In the future, it is expected that these approaches might be developed in line with the whole environmental assessment process, allowing the inclusion of: (i) a detailed planning of the different phases of the project; (ii) the definition of different mitigation scenarios for expected impacts and evaluation of their costs; and (iii) the establishment of limits to the level of impacts deemed acceptable.

Crucial assumptions regarding the success of an adaptive management programme for a given wind farm are:

- The programme should be dynamic, allowing for adjustment depending on the results obtained in the monitoring programmes and the ecosystem response to mitigation and compensation measures adopted at an early stage.
- The actions must be subject to the involvement of different stakeholders (developers, environmental experts, local authorities, civil groups etc.) in all the phases.

All of this information is critical since it significantly reduces the uncertainty associated with the project implementation and allows developers to assess their real social, economic and environmental viability.

References

APA. (2010). Guia Para Avaliação De Impacte Ambiental em Parques Eólicos [Methodological Guide to the Environmental Impact Assessment for Wind Farms]. Agência Portuguesa do Ambiente [Portuguese Environmental Agency]. Available from http://www.apambiente.pt/ (in Portuguese).

Arnett, E. B., Baerwald, EF., Mathews, F., Rodrigues, L., Rodriguez-Durán, A., Rydell, J., et al. (2015). Impacts of wind energy development on bats: A global perspective. In C. C. Voigt & T. Kingston (Eds.), *Bats in the anthropocene: Conservation of bats in a changing world* (pp. 295–323).

Arnett, E. B., Huso, M. M. P., Schirmacher, M. R., & Hayes, J. P. (2011). Altering turbine speed reduces bat mortality at wind-energy facilities. *Frontiers of Ecology and Environment, 9*(4), 209–214.

Arnett, E. B., & May, R. F. (2016). Mitigating wind energy impacts on wildlife: Approaches for multiple taxa. *Human-Wildlife Interactions, 10*(1), 28–41.

Baerwald, E. F., D'Amours, G. H., Klug, B. J., & Barclay, R. M. R. (2008). Barotrauma is a significant cause of bat fatalities at wind turbines. *Current Biology, 18*(16), R695–R696.

Barclay, R. M. R., Baereals, E. F., & Gruver, J. C. (2007). Variation in bat and bird fatalities at wind energy facilities: Assessing the effects of rotor size and tower height. *Canadian Journal of Zoology, 85,* 381–387.

BBOP. (2012). Standard on biodiversity offsets. *Business and Biodiversity Offsets Programme* (Washington, D.C.).

BBOP. (2013). *To no net loss and beyond: An overview of the business and biodiversity offsets programme* (p. 20). Washington, DC: Business and Biodiversity Offsets Programme.

Bennet, F., Culloch, R., & Tait, A. (2016). *Guidance on effective adaptive management and post-consent monitoring strategies.* Deliverables 5.2 & 5.4., RiCORE Project (p. 45).

BIO3. (2007). Monitorização da avifauna no Parque Eólico da Serra de Candeeiros—relatório 3. Bio3—Estudos e Projetos em Biologia e Valorização de Recursos Naturais, Lda., Almada.

BIO3. (2012). Monitorização da avifauna no Parque Eólico da Serra de Candeeiros—relatório 6. Bio3—Estudos e Projetos em Biologia e Valorização de Recursos Naturais, Lda., Almada.

BIOINSIGHT. (2016). Monitorização da avifauna no Parque Eólico da Serra de Candeeiros—relatório 9. Bio3—Estudos e Projetos em Biologia e Valorização de Recursos Naturais, Lda., Almada.

Birdlife International. (2015). *Review and guidance on use of "shutdown-on-demand" for wind turbines to conserve migrating soaring birds in the Rift valley/red Sea flyway.* Amman, Jordan: Regional Flyway Facility.

Brotas, G., Oliveira, JM., & Loureiro, C. (2015). Wind farms and a compensatory programme for the Iberian wolf in Portugal: Seven years of analysis. In J. Koppel & E. Schuster (Eds.), *Conference on wind energy and wildlife impacts: Book of abstracts.* Berlin, Germany.

Bulling, L., & Koppel, J. (2016) Exploring the trade-offs between wind energy and biodiversity conservation. In D. Geneletti (Ed.), *Biodiversity and ecosystem services in impact assessment.* Research Handbooks on Impact Assessment (pp. 299–320).

Cole, S. G., & Dahl, E. L. (2013). Compensating white-tailed eagle mortality at the smøla wind-power plant using electrocution prevention measures. *Wildlife Society Bulletin, 37,* 84–93.

Cordeiro, A., Mascarenhas, M., & Costa, H. (2013). Long term survey of wind farms impacts on Common Kestrel's populations and definition of an appropriate mitigation plan. In *Conference on Wind Power and Environmental Impacts, Stocolm, Sweden*

Drewitt, A. L., & Langston, R. H. W. (2006). Assessing the impacts of wind farms on birds. *Ibis, 148,* 29–42.

Drewitt, A. L., & Langston, R. H. W. (2008). Collision effects of wind-power generators and other obstacles on birds. *Annals of the New York Academy of Sciences, 1134,* 233–266.

EC. (2011a). Guidance on wind energy development in accordance with the EU nature legislation. Luxembourg: Office for official publications of the European Communities.
EC. (2011b). Wind energy developments and Natura 2000. EU Guidance on wind energy development in accordance with the EU nature legislation—European Commission, Brussels (p. 116).
Enevoldsen, P., & Sovacool, B. K. (2016). Examining the social acceptance of wind energy: Practical guidelines for onshore wind project development in France. *Renewable and Sustainable Energy Reviews, 53,* 178–184.
Gartman, V., Bulling, L., Dahmen, M., Geiβler, G., & Koppel, J. (2016a). Mitigation measures for wildlife in wind energy development, consolidating the state of knowledge—Part 1: Planning and sitting, construction. *Journal of environmental Assessment Policy and Management, 18*(3), 1650013.
Gartman, V., Bulling, L., Dahmen, M., Geiβler, G., & Koppel, J. (2016b). Mitigation measures for wildlife in wind energy Development, consolidating the state of knowledge—Part 2: operation, decommissioning. *Journal of environmental Assessment Policy and Management, 18*(3), 1650014.
GPWind. (2013). *Good practice guide—A new resource for reconciling wind energy development with environmental and community interests*. Available at https://ec.europa.eu/energy/intelligent/projects/sites/iee-projects/files/projects/documents/gpwind_good_practice_guide_gp_wind_en.pdf
Huso, M. M. P. (2010). An estimator of wildlife fatality from observed carcasses. *Environmetrics, 22,* 318–329.
IFC. (2012). *Performance standards on environmental and social sustainability* (p. 66). Washington: International Finance Corporation, The World Bank Group.
IUCN. (2016). Environmental & social management system (ESMS)—Standard on biodiversity conservation and sustainable use of natural resources. International Union for Conservation of Nature, available at https://www.iucn.org/sites/dev/files/iucn_esms_standard_biodiversity.pdf. Accessed in 2016–09–23.
Korner-Nievergelt, F., Korner-Nievergelt, P., Behr, O., Niermann, I., Brinkmann, R., & Hellriegel, B. (2011). A new method to determine bird and bat fatality at wind energy turbines from carcass searches. *Wildlife Biology, 17*, 350–363.
LEA. (2010). Monitorização dos efeitos da Medida de Minimização de Mortalidade do Parque Eólico do Outeiro. Relatório final. Laboratório de Ecologia Aplicada da Universidade de Trás-os-Montes e Alto Douro. Vila Real. Portugal, 78.
Marques, A. T. (2012). *Compensation and off-site mitigation strategies for endangered Bonelli's eagle populations in wind farms*. Workshop Wind Power and Biodiversity: Tools to measure, avoid and compensate impacts. EDP, organizado por Bio3, Fundación Global Nature e Global Nature Fund. 29 de Junho de 2012. Lisboa, Portugal.
Marques, A. T., Batalha, H., Rodrigues, S., Costa, H., Pereira, M. J. R., Fonseca, C., et al. (2014). Understanding bird collisions at wind farms: An updated review on the causes and possible mitigation strategies. *Biological Conservation, 179,* 40–52.
Marques, A. T., Paula, J., Pereira, M. J. R., Ramalho, R., & Rodrigues, S. (2015). Assessing the problem. In M. Mascarenhas, J. Bernardino, A. Paula, H. Costa, C. Bastos, A. Cordeiro, et al. (Eds.), *Biodiversity and wind energy: A bird's and bat's perspective* (pp. 30–51). Portugal: Bio3 and University of Aveiro.
Marques, J., Rodrigues, S., Paula, J., Mesquita, S., Cordeiro, A., Mascarenhas, M., et al. (2016). Mitigação de impactes dos parques eólicos em quirópteros minimizando perdas de produção: dois casos de estudo de adequação da velocidade de arranque das turbinas. CNAI'16. Évora. Portugal.
Marshall, R. (2001). Application of mitigation and its resolution within environmental impact assessment: an industrial perspective. *Impact Assessment and Project Appraisal, 19* (3), 195–204.

Martínez, J. E., Calvo, J. F., Martínez, J. A., Zuberogoitia, I., Cerezo, E., Manrique, J., et al. (2010). Potential impact of wind farms on territories of large eagles in southeastern Spain. *Biodiversity and Conservation, 19*, 3757–3767.

Moore, A. L., & McCarthy, M. A. (2010). On valuing information in adaptive-management models. *Conservation Biology, 24*(4), 984–993. doi:10.1111/j.1523-1739.2009.01443.x

Passos, I., Silva, M., Mesquita, S., Marques, A., Bernardino, J., Costa, H., et al. (2013). Aliens in wind farms—Preventing and monitoring impacts on vegetation. In *Communication presented at Conference on Wind Energy and Wildlife Impacts*. Stockholm, Sweden. Available from http://www.naturvardsverket.se/CWE2013/Emerging-issues/

Paula, J. J. S., Bispo, R. M. B., Leite, A. H., Pereira, P. G. S., Costa, H. M. R. G., Fonseca, C. M. M. S., et al. (2015a). Camera-trapping as a methodology to assess the persistence of wildlife carcasses resulting from collisions with human-made structures. *Wildlife research, 41*(8), 717–725.

Paula, A., Marques, J., Pereira, P. S., & Santos, J. (2015). Mitigation: A hierarchy of solutions. In M. Mascarenhas, J. Bernardino, A. Paula, H. Costa, C. Bastos, A. Cordeiro, et al. (Eds.), *Biodiversity and wind energy: A bird's and bat's perspective* (pp. 52–71). Portugal: Bio3 and University of Aveiro.

Pereira, M. J. R., Peste, F., Paula, A., Pereira, P., Bernardino, J., Vieira, J., et al. (2016). Managing coniferous production forests towards bat conservation. *Wildlife Research, 43*, 80–92.

Peste, F., Paula, A., da Silva, L. O., Bernardino, J., Pereira, P., Mascarenhas, M., et al. (2015). How to mitigate impacts of wind farms on bats? A review of potential conservation measures in the European context. *Environmental Impact Assessment Review, 51*, 10–22.

Santos, J., Cordeiro, A., & Costa, H. (2015). Reconciling wind and biodiversity: The integrated management approach. In M. Mascarenhas, J. Bernardino, A. Paula, H. Costa, C. Bastos, A. Cordeiro, et al. (Eds.), *Biodiversity and wind energy: A bird's and bat's prspective* (pp. 72–83). Portugal: Bio3 and University of Aveiro.

Santos, J., Marques, A T., Paula, A., Bernardino, J., Mascarenhas, M., & Costa, H. (2012). Compensation and off-site mitigation strategies for endangered Bonelli's eagle populations in wind farms. VIIth Polish Wind Energy Association (PWEA) Conference & Exhibition. Varsóvia, Polónia, May 22–23, 2012.

Stankey, G. H., Clark, R. N., & Bormann, B. T. (2005). Adaptive management of natural resources: Theory, concepts, and management institutions. General Technical Report. PNW-GTR-654. Portland, OR: U.S. Department of Agriculture, Forest Service, Pacific Northwest Research Station (p. 73).

Strickland, M. D., Arnett, E. B., Erickson, W. P., Johnson, D. H., Johnson, G. D., Morrison, M. L., et al. (2011). Comprehensive guide to studying wind energy/wildlife interactions. Washington, D.C., USA: Prepared for the National Wind Coordinating Collaborative.

Tomé, R. F., Canário, F., Leitão, A., Pires, N., Cardoso, P., & Repas, M. (2015). Radar assisted shutdown on demand ensures zero soaring bird mortality at a wind farm located in a migratory flyway. In J. Koppel & E. Schuster (Eds.), *Conference on wind energy and wildlife impacts: Book of abstracts*. Berlin, Germany.

Tomé, R., Leitão, A. H., Canário, F., & Pires, N. (2012). *Environmental management of wind farms—A successful onshore case study*. FAME international Workshop—Effects of marine renewables and other marine uses on biodiversity—Atlantic Area, Lisbon. Portugal.

USFWS. (2012). U.S Fish and Wildlife Service—Land-based wind energy guidelines, USA (p. 71).

van der BRUGGE, R., & van RAAK. (2007). Facing the adaptive management challenge: Insights from transition management. *Ecology and Society, 12*(2), 33. http://www.ecologyandsociety.org/vol12/iss2/art33. Accessed in 2014–03–11.

Walters, C. J., & Holling, C. S. (1990). Large-scale management experiments and learning by doing. *Ecology, 71*,2060–2068.

Westgate, M. J., Likens, G. E., & Lindenmayer, D. B. (2013). Adaptive management of biological systems: A review. *Biological Conservation, 158*, 128–139.

Williams, B. K., & Brown, E. D. (2012). *Adaptive management: The U.S. department of the interior applications guide*. Adaptive Management Working Group, U.S. Department of the Interior, Washington, DC.
Williams, B. K., & Brown, E. D. (2014). Adaptive management: From more talk to real action. *Environmental Management, 53,* 465–479. doi:10.1007/s00267-013-0205-7.
Williams, B. K., & Brown, E. D. (2016). Technical challenges in the application of adaptive management. *Biological Conservation, 195,* 255–263.
Williams, B. K., & Johnson, F. A. (2013). Confronting dynamics and uncertainty in optimal decision making for conservation. *Environmental Research letters, 8,* 025004.
Williams, B. K., Szaro, R. C., & Shapiro, C. D. (2007). *Adaptive Management: The U.S. Department of the Interior Technical Guide*. Washington, DC: Adaptative Management Working Group. U.S. Departmente of the Interior.
Williams, B. K., Szaro, R. C., & Shapiro, C. D. (2009). *Adaptive management: The U.S. Department of the interior technical guide*. Washington, DC: Adaptive Management Working Group, U.S. Department of the Interior.

Index

M. Mascarenhas et al. (eds.), *Biodiversity and Wind Farms in Portugal*,
DOI 10.1007/978-3-319-60351-3

MIX
Papier aus verantwortungsvollen Quellen
Paper from responsible sources
FSC® C105338

If you have any concerns about our products,
you can contact us on
ProductSafety@springernature.com

In case Publisher is established outside the EU,
the EU authorized representative is:
Springer Nature Customer Service Center GmbH
Europaplatz 3, 69115 Heidelberg, Germany

Printed by Libri Plureos GmbH
in Hamburg, Germany